Six Septembers: Mathematics for the Humanist

Patrick Juola and Stephen Ramsay

Zea Books
Lincoln, Nebraska
2017

Version March 27, 2017

ISBN 978-1-60962-111-7

https://doi.org/10.13014/K2D21VHX

Zea Books are published by the University of Nebraska–Lincoln Libraries
Electronic (pdf) edition available online at http://digitalcommons.unl.edu/zeabook/
Print edition available from http://www.lulu.com/spotlight/unlib

University of Nebraska Lincoln

Contents

Introduction

> Look at mathematics: it's not
> a science, it's a monster
> slang, it's nomadic.
>
> Deleuze and Guattari
> *A Thousand Plateaus*

This is a book on advanced mathematics intended for humanists.

We could defend ourselves in the usual way; everyone can benefit from knowing something about mathematics in the same way that everyone can benefit from knowing something about Russian literature, or biology, or agronomy. In the current university climate, with its deep attachment to "interdisciplinarity" (however vaguely defined), such arguments might even seem settled.

But we have in mind a much more practical situation that now arises with great regularity. Increasingly, humanist scholars of all stripes are turning their attention toward materials that represent enormous opportunities for the future of humanistic inquiry. We are speaking, of course, of the vast array of digital materials that now present themselves as objects of study: Google Books (rumored to be over twenty *million* volumes at the time of this writing);

vast stores of GIS data; digitized census records; image archives of startling breadth; and above all, the World Wide Web itself, which may well be the most important "human record" of the last five hundred years.

We have great confidence that the humanities will not only benefit from the digital turn, but that it will develop its own distinct methodologies for dealing with this sudden embarrassment of riches. At the same time, the nature of that material has created a world in which English professors are engaged in data mining, historians are becoming interested in ever more advanced forms of regression analysis, librarians are learning to program, art historians are exploring image segmentation, and philosophers are delving into the world of computational linguistics.

Or rather, some are. The field now known as "digital humanities"—a strikingly interdisciplinary (and international) group of scholars engaged in such pursuits—remains a comparatively small band of devoted investigators, when compared to the larger and older disciplines from which they come. The reasons for this should be obvious; all of the techniques mentioned above require that one stray into intellectual territory that was until recently the exclusive province of the sciences (whether computer, social, or "hard").

The humanist scholar who undertakes to study data mining, advanced statistics, or text processing algorithms often approaches these subjects with great optimism. After many years spent studying subjects of comparable complexity in literature, history, or ancient languages, he or she certainly has reason to believe that whatever knowledge is demanded will become tractable with the right amount of determination and hard work. Data mining is complicated, but so is

French literary theory. Everyone has to start somewhere.

But that attempt often ends badly. A humanist scholar who tries to read an article in, say, a computer science journal—or even, for that matter, an introductory text on a highly technical subject—will quite often be brought up short the minute things get mathematical (as they almost inevitably will). There is much that the assiduous student can glean with the proper amount of effort in such situations, but when the equations appear, old phobias appear with them. It might as well be Greek. It would be *easier* if it were Greek.

We have very particular ideas about what is lacking here. In our experience, the problem stems from two deficiencies:

1. Humanists without training in the formal notation of mathematics—notation which in nearly every case values concision over explanatory transparency—cannot make their way through an argument that depends on that notation. And there is, in general, no way to "figure it out" without that training.

2. Humanists often lack the proper set of concepts for dealing with mathematical material—concepts that are independent of their specific manifestation in equations and proofs, but without which mathematical arguments become mostly unintelligible.

This book does not explain all of the mathematics that anyone is likely to encounter in the humanities. That would be impossible, to begin with; part of what makes the digital humanities so exciting is the constant eruption of lateral thinking that allows a scholar to see some tool or technique, apparently unrelated to the humanities, as a new vector for study and contemplation. Our purpose, instead, is to impart the *concepts* that we believe underlie most of the mathematics that you are likely to encounter, and to unfold the notation in a way that removes that particular barrier completely. This book is, in other words, a primer—a book that you can use to develop the skills and habits of mind that will allow you take on more complicated technical material with confidence.

Some of what we talk about in this book is directly applicable to problems in the humanities, but that is not our main concern. Much of it is devoted to material that might never appear in the ordinary course of "doing" digital humanities, sociology, game studies, or computational linguistics. Yet we firmly believe that all of what we discuss here is in the back of the mind of anyone who does serious technical work in these areas. This book, to put it plainly, is concerned with the things that the author of a technical article knows, but isn't saying. Like any field, mathematics operates under a regime of shared assumptions, and it is our purpose to elucidate some of those assumptions for the newcomer.

The individual subjects we tackle are (in order): logic and proof, discrete mathematics, abstract algebra, probability and statistics, calculus, and differential equations. This is not at all the order in which these subjects are usually taught in school curricula, and indeed, it is possible to take a course of study that does not include all of them. Our ordering is borne of our own sense of how best to convey the concepts of mathematics to humanists, and is, like mathematics itself, strongly cumulative. We have made no attempt to write chapters that can stand on their own, and would therefore strongly suggest that they be read in order.

Of course, any of these areas could be the subject of a multi-volume series of books. In fact, all of them are, and we recommend additional reading where appropriate. Our aim here is to introduce the major concepts and terminology of the areas, not to provide a detailed exploration of the technical foundations or even necessarily to catalog the major results in the areas. Instead, we present concepts and vocabulary appropriate to the first few weeks of—a September, so to speak—of a rigorous, university-level course, but (we hope) with a presentation more focused on understanding and application and less on proof techniques.

We assume knowledge of nothing more than basic algebra (along with, perhaps, elementary geometry and a general sense of the main ideas in trigonometry). We are aware, of course, that for most of our readers, decades have passed since these subjects were first unfolded. We provide a bracingly brief refresher course in algebra in the appendix, though you may find it helpful to turn to it only when needed.

This book is written by two humanists. One is a professor of English who found himself drawn to the digital humanities, and then struggled mightily (for years) through the very situation we describe above before gaining facility with the subject. The other is a professor of mathematics and computer science who has spent his entire career in the company of humanists, struggling to understand their questions and their methodologies. This collaboration has been sustained at every turn by certain attitudes which hold in common, and which we have tried to bring to these pages.

First, we have nothing but contempt for the phrase "math for poets"—a sobriquet we consider only slightly less demeaning and derisive than "logic for girls." The implication that this idea ensconces (often, sadly, in course titles) is one that we reject, because it implies that the subject can only be understood if certain, largely negative things are assumed about the student's abilities. We assume that our readers are motivated adults who need explanations, not some radically reframed version of the subject that makes rude assumptions about "the way they think." We have tried very hard to explain things as clearly as we can, but we do not shy away from "real" mathematics. Since it is an introductory text, we are aware not only of roads not taken, but of simplifications that scarcely convey the depths of the subjects we're discussing. But again, we have taken this route in order to lay proper emphasis on concepts, not to present a watered-down version of a complicated subject. The "for" in our title is as it appears in the phrase, "We have a gift for you."

We also believe firmly in autodidacticism. Aside from having a very noble tradition in mathematics (some of the greatest mathematicians have been, properly speaking, self-educated amateurs), this mode of learning becomes a necessity for those of us long out school and without opportunity to undertake years of organized study. At the same time, we believe in community. We aren't so bold as to think that we've hit upon the most lucid explanations possible in every case, and we encourage you to seek out mentors who can help. We even provide a forum for such discussions on the website that accompanies this book.

Finally, we believe in the humanities. Both of us have an enduring fascination with mathematics, but our passions as scholars remain focused on the ways in which this subject holds out the possibility of fruitful interaction with the study of the human record. The hope we have for this book is not that it will create new mathematicians, but that it will embolden people to see new possibilities for the subjects to which they've devoted themselves.

To Deliver You from the Preliminary Terrors

Having just extolled your virtues as a serious, motivated, autodidact, we nonetheless need to admit that reading mathematics can be a difficult matter—even in the context of an introductory text like this.

We borrow the title of this section from a similar note to the reader in Silvanus P. Thompson's 1910 book *Calculus Made Easy*—a widely acknowledged masterpiece of clear and elegant mathematical exposition that has, in many ways, inspired our own work. We especially like the subtitle: *Being a Very-Simplest Introduction to Those Beautiful Methods of Reckoning Which Are Generally Called by the Terrifying Names of the Differential Calculus and the Integral Calculus.* We are not so far from being beginners in this subject as to have no sense of the terror to which Thompson refers; it visits everyone, sooner or later.

You will almost certainly say to yourself at some point: "Wait. I am lost." We believe firmly that there are several ways out of this pathless wood—the chief one being the avoidance of those conditions that get us there in the first place. In particular:

1. We who are used to reading books and scholarly articles in the

humanities seldom get "stuck" when reading something. If we miss an idea, we can usually recover as the narrative proceeds. Mathematical reading is not like this. The subject is by nature cumulative, and some of the apparently trivial things we say at the beginnings of chapters have great moment later on. It is therefore necessary to check your understanding constantly: "Did I get that? Do I really understand that?" We really can't emphasize this enough.

2. As a direct corollary: Mathematical reading is very slow. It is not at all uncommon for a professional mathematician to spend weeks "reading" an eight-page article. We doubt that anything in our pages will require such patience, but we want to caution against too brisk a pace and assure you that even the most gifted mathematician is moving at a fraction of the speed with which an ordinary reader moves through, say, a novel.

3. "Mathematics," to quote one of our beloved instructors, "is not a spectator sport." We have no desire to make our subject feel like homework, but there's no getting around the fact that to understand mathematical ideas, you have to *do* mathematics. We therefore humbly suggest that you seek out problems to solve. Nothing will elucidate the subject faster than trying to work out how a given concept applies to a given problem. We provide some sample exercises on the website that accompanies this book, but the best problems are those that occur to you as you're working in some other area (say, looking at the mechanics of a game or trying to understand a social network).

> Our emphasis, once again, is on concepts, but many of these same concepts emerged in the context of real-world problems to be solved. There's no substitute for re-creating the conditions under which those insights first became manifest.

We are teachers, and we suspect that many of our readers are as well. Even if you are not a teacher, you can easily understand something that every experienced teacher knows very well: that when it comes to teaching complicated subjects, half the battle is getting students to recognize themselves as the sort of people who can do it well. We have to get to "I got it" early and often in order to build the confidence that is required to proceed.

The temptation, when one becomes lost, is to assume that it is your fault—that you're just not built for this. We think that many things may have gone wrong in such cases (the authors' lack of clarity among them). But we think (and there is research to prove it) that human beings are built for mathematical thinking in the same way that human beings are built for speaking and using tools. When the moment comes (and it will), our advice is to heed the points above and to recognize that people with significantly fewer intellectual gifts than yourself have successfully mastered this material. Thompson offers an ancient proverb as the epigraph to his text that is beloved by scientists, engineers, and mathematicians the world over, and which we think is a good thing to repeat during the dark night: *What one fool can do, another can.*

1 Logic and Proof

The great twentieth-century British mathematician, G. H. Hardy (1877–1947), once wrote, "I have never done anything 'useful.' No discovery of mine has made, or is likely to make, directly or indirectly, for good or ill, the least difference to the amenity of the world" [5]. Yet he did allow that one of his discoveries had indeed been important—namely, his "discovery" of the great Indian mathematician Srinivasa Ramanujan (1887–1920). Ramanujan, with almost no training in advanced mathematics, had managed to derive some of the most important mathematical results of the last five centuries while working as a clerk at the Account-General's office in Madras. He had sent some of his work to several members of the mathematics faculty at Cambridge, but only Hardy had recognized Ramanujan as a genius. They would become lifelong friends and collaborators.

For Hardy, the beauty of mathematics was to be found in the layered elegance by which theorems are established from more elementary statements. Ramanujan, by contrast, seems to have found the tedium of proof too burdensome. "It was goddess Namagiri, he would tell his friends, to whom he owed his mathematical gifts. Namagiri would write the equations on his tongue. Namagiri would bestow mathematical insights in his dreams" [8]:

> Hardy used to visit him, as he lay dying in hospital at Putney. It was on one of those visits that there happened the incident of the taxi-cab number. Hardy had gone out to Putney by taxi, as usual his chosen method of conveyance. He went into the room where Ramanujan was lying. Hardy, always inept about introducing a conversation, said, probably without a greeting, and certainly as his first remark: "I thought the number of my taxi-cab was 1729. It seemed to me rather a dull number." To which Ramanujan replied: "No, Hardy! No, Hardy! It is a very interesting number. It is the smallest number expressible as the sum of two cubes in two different ways." [5]

For Hardy, the story was undoubtedly meant to stand as a tribute to his friend. For surely, only someone with a very special relationship with numbers could possibly make such an observation spontaneously. But for all that, one can almost imagine Hardy's thoughts on the ride home. Is it true? And is it, in fact, interesting?

We can settle the first question with a **proof**—a demonstration that the statement must be true (or that it must be false). To do this, we would have to show that 1729 is in fact the sum of two cubes, that there are exactly two such cubes that sum to 1729, and that no smaller number can be created this way. The second question is undoubtedly a more subjective one, though we might discover that the proof itself reveals something else about mathematics, or that the result can be used in other, perhaps more "interesting" demonstrations.

We have, then, a **conjecture**: 1729 is the smallest number expressible as the sum of two cubes in two different ways. Having proved it, it would become a **theorem**. If that theorem could in turn be used to establish further theorems, it would become a **lemma**. The study of those methods and processes by which arguments are made, theorems established, and lemmas formed is called **mathematical logic**.

The idea of proving ever-more-elaborate statements using the results of earlier proofs is familiar to most people from secondary-school geometry. Euclidean geometry, in fact, provides an excellent introduction not only to logic, but to the place of logic within mathematics. Students who enjoy the subject will remark on its beauty—on the way that things "fit together." Yet others are quick to point out that there is something fundamentally wrong with a system that has to define basic things like the notion of a point as (to quote Euclid) "that which has no part" and a line as a "breadthless length." [4] Even the most enthusiastic student will at some point ask why it is necessary to prove so many things that are obviously true at a glance.

Such questions hint at problems and conundrums that are among the oldest questions in philosophy. At stake is not merely the virtues and limitations of this or that system, but (one suspects) the virtues and limitations of our ability to reason about systems in general. Behind the many methodologies (or "logics") that have been proposed is a set of basic questions about methodology itself, and about the nature of those epistemological tools that we have traditionally employed to establish truth and reject falsehood in matters ranging from the existence of triangles to the proper form of government.

Since it is the purpose of this book to present mathematics to humanists, we will focus here on the tools themselves. Our intention, however, is not to put these larger questions into abeyance, but rather to draw them forth more forcefully by showing how these systems actually work and what it is like to work with them. Toward the end, we revisit some of the major philosophical insights concerning logic and proof for which twentieth-century mathematics may be most well remembered.

1.1 Formal Logic

Logic usually traffics in the truth and falsehood of statements. To say this is already to restrict our view, since not all statements are susceptible to such easy classification. Some statements appear to have nothing at all to do with truth and falsehood ("Look out for that tree!"); others might have something to do with truth and falsehood, but are not themselves true or false ("Was that tree there last week?").

In technical terms, a statement that can be true or false is said to be a **proposition** (or to have **propositional content**). This puts us immediately in more broadly philosophical territory, since some would like to declare a **law of the excluded middle** and say that mathematics is exclusively concerned with statements that are either true or false. This might seem an undue restriction; surely a system constrained in this way is already barred from the investigation of more nuanced problems. Yet propositions hold out a possibility that logicians have been unable to resist—namely, the idea that a statement might be "analytically" true or true by virtue of its structure. For this reason, one might view logic as the study of the structure of arguments. The goal is not so much to prove this or that, but rather to say that statements of a certain form—independent even of their content—must always serve to support a given argument.

1.1.1 Deduction and Induction

Though there are many logical systems, mathematical logicians traditionally distinguish between the two broad classes familiar to us from philosophy and science: deductive and inductive. **Deductive logic**, you'll recall, is the process of making inferences from general premises to specific conclusions. If, for example, we know that (i.e. it is true that) all fish live in the water, then any specific fish—say, this particular goldfish—must itself live in the water. By contrast, **inductive logic** is the process of reasoning from specific instances to general conclusions. Every fish I've seen lives in the water, so therefore I conclude that all fish live in water.

This common distinction hints at the idea of analytical truth. Arguments based on deductive reasoning are guaranteed to be correct (provided we follow certain rules), while those based on inductive reasoning are not. So even if all the swans I have ever seen are white, the statement "All swans are white" can be demonstrated as false the minute an informed reader demonstrates the existence of black ones. Some logicians have suggested that this guarantee is a better criterion for defining the difference between deductive and inductive logic. It is, after all, possible to reason to some kinds of general statements from specific examples. Seeing a goldfish living in a fishbowl, I can say with certainty that the statement "all fish live in the ocean" cannot be true, but under the revised criterion, this would be considered a deductive argument.

1.1.2 Validity, Fallacy, and Paradox

Validity

But how can we tell if we are reasoning "correctly?" Or, to put it another way: How can we tell that certain arguments support their conclusions and that others do not? From a mathematical perspective, the first step toward answering questions like these necessitates careful and precise definition of the terms. Defining things doesn't itself guarantee correctness (we could, after all, define things incorrectly), but definitions increase the likelihood that the operative assumptions are shared ones. The old Latin legal term *arguendo* ("for the sake of argument") captures the sense in which definitions are put forth in mathematics.

For the sake of arguing about mathematics, then, we define a few words more restrictively than we do normally, and say that an **argument** is a series of propositions—which is to say, a series of statements that are either true or false. Propositions that the speaker believes to be true (or accepts provisionally as true) and which are offered for your acceptance or belief are called **premises**. The last proposition is conventionally called the **conclusion**. The tantalizing possibility, as we mentioned before, is that this *structure* has some kind of inevitability to it. Because the premises are true, the conclusion *must* be true.

We can structure and formalize our earlier argument as follows:

$$\text{(First premise)A goldfish is a fish.} \tag{1.1}$$
$$\text{(Second premise)All fish live in water.} \tag{1.2}$$
$$\text{(Conclusion)} \therefore \text{Goldfish live in water.} \tag{1.3}$$

1

It should be intuitively obvious[2] that the premises of this argument offer support for the conclusion, while the premises of the following argument do not:

$$\text{(First premise) My friend's cat is a Siamese cat.} \tag{1.4}$$
$$\text{(Second premise) Some Siamese cats have blue eyes.} \tag{1.5}$$
$$\text{(Conclusion)} \therefore \text{My friend's cat has blue eyes.} \tag{1.6}$$

Such examples naturally seem almost insulting in their simplicity, but they go to the heart of the matter. Saying that something is "intuitively obvious" doesn't guarantee that it is true. Such statements

[1] ∴ is just a conventional symbol to mark the conclusion of an argument; when you're reading it, you just say "therefore."

[2] "Intuitively obvious" is among the most notorious weasel-phraess in mathematics; in some contexts, this has been called "proof by intimidation." At its best, it means that the statement under discussion should, in fact, be seen to be true based on common sense. At its worst, it means that the writer thinks something is true, but has no idea how to show it and so is trying hard to pretend that the Emperor is well-dressed. We'll use it (or something like it) ourselves from time to time; be sure to distrust us every time we do).

criticize our intuition (which might be another term for our experiential beliefs) in advance, in the hope that this appeal will bias us in favor of agreement. A great deal of substantive arguing in the real world depends on such appeals, and it is not at all cogent to say that such rhetorical moves are illegitimate. In the context of mathematical logic, however, it seems an unlikely path toward "structural truth."

Unfortunately, at this point in our exposition, we don't have anything better. Informally, of course, the problem appears to lie with the word "some." If "some" Siamese have blue eyes, some of them might not have blue eyes, and we don't know which type my friend has.

A **valid** argument is therefore traditionally defined as an argument where the truth of the premises entails the conclusion (that is, it is not possible for the premises to be true while the conclusion is false). Being rigorists by nature, logicians go further and say that these conditions must obtain not only in the world, but in "all possible worlds." This latter stipulation attempts to move the question entirely beyond intuition and experience, even to the point of imagining radically altered circumstances in which the argument might not be valid. The argument about Siamese cats is **invalid**, because (irrespective of the actual cat) my friend's cat might have green eyes. But twist and turn as we might, we can't invalidate the first argument.

Declaring an argument invalid doesn't necessarily mean that the conclusions are wrong. The following argument is identical in structure to the second (lines 1.4–1.6), and yet the conclusion is still true:

Paul Newman is a person.	(true)	(1.7)
Some people have blue eyes.	(true)	(1.8)
∴ Paul Newman has blue eyes.	(true)	(1.9)

It is invalid, not because it reaches a false statement, but because identically structured arguments *can* lead to false conclusions.

Similarly, an argument can be valid without its conclusions being true:

Paul Newman is a person.	(true)	(1.10)
All people have seven eyes.	(false)	(1.11)
∴ Paul Newman has seven eyes.	(false)	(1.12)

In this case, we can see that at least one of the premises is false. Yet we can (with some effort) imagine a possible world where all people do have seven eyes and where Paul Newman is considered a person. On such a planet, Paul Newman would also have seven eyes.

A valid argument can be thought of as as a truth-making machine; if you put true propositions into it, the propositions you get out of it are guaranteed to be true. You can use this machine either to generate new truths, or to check (prove) that statements you already believe are in fact true.

Fallacy

Mathematical logic is noticeably strict when it comes to valid arguments; the rest of the world is far less so. Many highly persuasive arguments made in the context of business or politics (or even just ordinary conversation) fail the "possible world" criterion. Most commonly accepted scientific "truths" would likewise fail due to reliance on inductive (and therefore invalid) argument. Statistics, too, is mostly about assessing the degree of support offered by evidentiary structures that are formally invalid (again, because they are mostly inductive). Yet invalid arguments are often used illegitimately to sell things, propagandize, or simply to lie.

The tradition of compiling lists of **fallacies**—relatively convincing, but invalid arguments—is an ancient one. In some cases, the invalidity is rather obvious, as it is with the so called *argumentum ad baculum* (literally, "appeal to the stick"): "If you don't believe this proposition, I will hit you with this stick." But other fallacies are far more subtle, and provide useful illustration of the idea of argumentative structure.

The two valid arguments presented above both have the same general form. We can illustrate this by replacing phrases like "fish" or "people" with letters as placeholders, yielding the following structure:[3]

[3]It may seem strange, at least from a grammatical standpoint, to turn "Paul Newman is a person" into "All S are M," but the latter captures a wide range of expressions in natural language. In this case, we are restructuring the English statement into the rather stilted, "All things that are Paul Newman are things that are people."

All S are M. (1.13)

All M are P. (1.14)

∴ All S are P. (1.15)

Suppose that one were to swap the terms of the second premise, like so:

All S are M. (1.16)

All P are M. (1.17)

∴ All S are P. (1.18)

The argument now has a pleasing symmetry, and yet it is jarringly invalid, as the following example makes clear:

All money [in the United States] is green. (1.19)

All leaves are green. (1.20)

∴ All money is leaves (and thus grows on trees). (1.21)

This fallacy is called the **fallacy of the undistributed middle**, and the symbolic version above illustrates why. An argument with an "undistributed middle" has the middle term—the one that does not appear in the conclusion—appearing only on the right side (or left side) of the sentence (the predicate). Once again, we can see that it is invalid without even knowing what S, M, and P are.

This kind of structure is often used in arguments to support the idea of guilt by association. For example, if all Communists support socialized medicine, and a candidate's political opponent also supports socialized medicine, the candidate could use this in their campaign literature to "support" the argument that his or her opponent is a closet Communist. Yet even here, there are instances where the fully fallacious argument can be reasonably persuasive. For example, a prosecuting attorney might find it helpful to point out that the murderer had a certain type of DNA that matches the defendant, as evidence of the defendant's guilt.

A full catalog of fallacies with accompanying illustration would be a full book in its own right and would take this chapter too far afield. However, we'd like to mention one fallacy that is considered extremely dangerous in mathematical logic: namely, the so-called **fallacy of equivocation**, which occurs when the meaning of a term shifts in the middle of an argument.

$$\text{A feather is light.} \tag{1.22}$$

$$\text{What is light cannot be dark.} \tag{1.23}$$

$$\therefore \text{A feather cannot be dark.} \tag{1.24}$$

Much of mathematical discourse is centered around exact definitions and their implications to prevent accidental ambiguity for this very reason.

Paradox

What of sentences that are neither true nor false? As we've already mentioned, some kinds of speech acts don't have propositional content—questions, exclamations, orders, and so forth. Yet a broad category of sentences appear superficially to be propositions, but upon closer inspection cannot be. Such sentences are called **paradoxes**.

A classic example of paradox is the **Barber Paradox**. "*In a small town with only one barber, the barber shaves all and only the men who are not self-shavers (i.e. do not shave themselves). Is the barber a self-shaver?*" If the barber is not a self-shaver, then he must shave himself (because he shaves all the men who aren't self-shavers), and therefore he must be a self-shaver. Conversely, if he is a self-shaver, then (since he shaves only men who aren't self-shavers), he must not shave himself, and therefore is not a self-shaver.[4]

The paradox, then, is that if the sentence "The barber is a self-shaver" is true, then the same sentence must be false. If the sentence is false, then it has to be true. But it can't be both true and false (the law of the excluded middle again), and so it therefore can't be either true or false without contradicting itself. It is a paradox.

In practical terms, a paradox means that there is something wrong with your assumptions and definitions. In the twenty-first century,

[4]We can also declare that the barber cannot have a beard. In fact, we can prove that no one in this town has a beard. A man with a beard demonstrably does not shave himself, and therefore must be shaved by the barber, according to the strict terms of the problem statement. But if he's shaved by the barber, he doesn't have a beard.

this paradox loses something of its punch when we note that female barbers exist. If you assume (implicitly) that the barber is a man, the paradox is genuine. However, if the barber is a woman, then the situation can be resolved by noting that the problem specifically states that she shaves "all and only the men"—i.e. the problem is explicit about the fact that her client list is exclusively male. Naturally, then, she doesn't shave herself, but there is no paradox. She simply doesn't shave herself—if she is shaved at all, someone else must do it for her.

Paradoxes have been hugely influential in mathematics precisely because they are so good at rooting out hidden assumptions and problems with definitions. Obviously, any logical scheme that permits the inference of paradoxes is to be avoided. As we'll see later, one of the most earth-shattering insights in modern mathematics relies on the elucidation of a paradox.

1.2 Types of Deductive Logic

So in order to be able to analyze deductive arguments, it becomes necessary to represent them accurately and concisely. This section presents several different systems of representation and analysis, ranging from ancient Greek logic to the present day. We also present some of the notation of modern logic in common use in the literature of both mathematics and analytical philosophy.

1.2.1 Syllogistic Logic

Syllogistic logic, the subject of a collection of treatises by Aristotle, is among the most enduring and influential forms of logic. During the period that stretches from the high middle ages to the nineteenth century, nearly every area of intellectual endeavor in the West—from mathematics to theology—became heavily influenced by the basic terms and "cognitive style" of syllogistic logic.[5] Even as late as 1781, Immanuel Kant—hardly an undiscerning thinker—could innocently declare that "since Aristotle, [logic has] been unable to take a single step forward, and therefore seems to all appearance to be finished and complete" [9].

A **syllogism** (from a Greek word meaning "inference" or "conclusion") is the sort of argument we've been examining up until now: an argument based on two premises and a conclusion. The ancients discerned four basic forms that the statements of a syllogism could take:

A: All A is B. (universal affirmative)

E: No A is B. (universal negative)

I: Some A is B. (particular affirmative)

[5]We speak of "the West" in order to distinguish a particular form of logic that developed alongside similar systems in India and within the broad sweep of Islamic philosophy. The Aristotelian influence on Islamic thought, and the influence of thinkers like Avicenna and Averroes on European mathematics, is well beyond the scope of this book, but is an important part of the story of the broader story of modern mathematics.

O: Some A is not-B. (particular negative)

The following sentences are all examples of A-type propositions:

- All cats are mammals.
- All basketball players are tall (i.e., are tall-things).
- All insomniacs snore
  (i.e., All things-that-are-insomniacs are things-that-snore).
- Bill Gates is rich
  (i.e., All things-that-are-Bill-Gates are things-that-are-rich).

The following sentences would be I-type propositions:

- Some cats are black (we don't specify which).
- One of my nieces is a champion speller.
- There are black swans in Australia (i.e., Some swans are black-things).

This is an O-type proposition:

- Not all lawyers are crooks

. . . and this is an E-type proposition:

- None but the brave deserve the fair
  (i.e. No things-that-are-not-brave are things-that-deserve-the-fair)

With only four different sentence types, and three sentences in each syllogism, there are fewer than 200 different possible syllogisms, only some of which—fewer than 20—are valid. It is therefore possible to compile a table or list of all the valid syllogisms. Such lists were compiled during the Middle Ages, and various syllogisms given individual names. For example, the **Barbara syllogism** we discussed above is made up of three A-type sentences:[6]

Major premise-A	My sister's goldfish is a fish.	(1.25)
Minor premise-A	All fish live in water.	(1.26)
Conclusion-A	∴ my sister's goldfish lives in water.	(1.27)

Similarly, the **Darii syllogism** has an A-type major premise, an I-type minor premise, and an I-type conclusion:

Major premise-A	All trout are fish.	(1.28)
Minor premise-I	No fish live in trees.	(1.29)
Conclusion-I	∴ No trout live in trees.	(1.30)

Enumerating all of the valid syllogisms was considered a good way of learning which ones were and were not "valid" arguments within this framework. A more powerful property emerged, however, when the valid syllogisms were rewritten as prescriptive rules

[6]The designations "A," "E," "I," and "O" were assigned by teachers of logic in the middle ages, and have persisted in the mnemonics used for the various syllogisms: Barbara (bArbArA), Festino (fEstInO), Cesaro (cEsArO), etc.

about how to reason. For example, we can restructure the Barbara syllogism as:

> If you have a pair of premises "All A are B" and "All B are C," you may infer "All A are C" (without regard to the exact terms of A, B, and C).

This is an example of an **inference rule**. Such a rule can be used in one of two ways. First, it can be used, once again, as a truth-making machine. Given a pair of statements in the appropriate form, the "machine" can infer the appropriate conclusion. It can also be used, in reverse, as a system for checking the truth of a statement or the validity of an argument. If a statement is offered as a conclusion, we could check to see whether the pair of premises corresponds to a valid syllogistic structure. This checking can be done without detailed knowledge of the propositions in question; in fact, if the statements are offered in a sufficiently stereotypical form, it could be done by a computer with nothing more than text matching.

While the influence of syllogistic logic is everywhere evident in the systems which came to replace it, it has now more-or-less disappeared as a subject of intensive inquiry in philosophy and mathematics. That is due, at least in part, to evolving notions of what logic itself is about. Earlier, we described logic as a system for discerning the truth or falsehood of statements. While that is true, we can also think of logic as an attempt to provide rigorously defined meanings and interpretations for particular words. Syllogistic logic provides quite precise descriptions for words like "all," "some," and "none." It falters, however, even with the introduction of a word like "or":

$$\text{Either Timothy or Sasha is a cat.} \tag{1.31}$$
$$\text{Timothy is not a cat.} \tag{1.32}$$
$$\text{Sasha is a cat.} \tag{1.33}$$

This resembles a syllogism, but there is no way within the framework to get from a statement about some of the members of a group (the set of things that are either Timothy or Sasha) to a statement about the individuals in that group.

In general, the problem with syllogistic logic is not so much a matter of logical validity, but of expressive power. The revolution in logic that began in the nineteenth-century with the work of people like George Boole (1815–1864) and Augustus de Morgan (1806–1871) was occasioned not only by the quest for better "truth machines," but by attempts to understand language and its relation to thought and meaning. As we shall see, it also led to revolutionary ideas about the foundations of mathematics itself.

1.2.2 Propositional Logic

The first development toward a more expressive logic moved the focus of concern from the terms of a proposition to the entire proposition taken as a whole. Consider the following:

- "Percy is a cat"
- "Quail are a kind of lizard"

- "Rabelais was the author of both *Gargantua* and *Pantagruel*"
- "Sasha is a cat"
- "Timothy is a chinchilla"

These are, according to our earlier definition, propositions, in the sense that they can be either true of false. We can begin the construction of a **propositional logic** by attempting to discern the structures that govern the relationships among such propositions.

We can begin with the observation that to every proposition, there is a contrary proposition, one that is false if the original is true, and true if the original is false. Linguistically, we can construct such propositions by prepending the phrase "It is not the case that ..." to the beginning of any proposition. Thus:

- "It is not the case that Percy is a cat"
- "It is not the case that quail are a kind of lizard"
- "It is not the case that Rabelais was the author of both *Gargantua* and *Pantagruel*"
- "It is not the case that Sasha is a cat"
- "It is not the case that Timothy is a chinchilla"

These are simple propositions, but one of them is not as simple as it might be. The Rabelais sentence really indicates both that

- Rabelais is the author of *Gargantua.*

and that

- Rabelais is the author of *Pantagruel.*

Joining these elementary propositions together with "and" creates a single (linguistic) statement that asserts an underlying combination of the two (logical) propositions. Clearly, we could join the elementary propositions differently by saying, for example, "Rabelais is the author of *Gargantua* but not the author of *The Decameron.*" In either case, however, it becomes possible to speak of the connecting term as having a logic unto itself.

Let's say I assert the following:

- Sasha is a cat and Timothy is a chinchilla.

Under what circumstances would we regard this statement as false? Clearly, it would be false if Sasha was a dog and Timothy a parakeet, but we would also consider it false if only one of those were true. That is, if Sasha is indeed a cat but Timothy is a parakeet, we would consider the entire proposition to be false.

Other connectives can have different logics. In the case of a **disjunction** (or), both "disjuncts" would have to be false for the compound proposition to be false.

- Sasha is a cat or Timothy is a chinchilla

Propositions embedded within "if. . . then" express the concept of **implication**: "If Rabelais was the author of *Gargantua,* then Rabelais was (also) the author of *Pantagruel.*" Despite the apparent simplicity of such phrases, the precise logic of such connectives can be more difficult to grasp. Our intuitive understanding is that the statement "if the Yankees win two more games, they will make the playoffs" has two main implications. First, if the statement is true and the Yankees win two more games, they will indeed make the playoffs. Second, if the Yankees win two more games, but don't make the playoffs, then the statement was false. But what about the case where the Yankees don't win? If the Yankees don't win two more games, but slip into the playoffs anyway, was the original statement false? How about if the Yankees don't win any more games, and don't make the playoffs? Was the speaker lying? Intuition suggests, and propositional logic agrees, that an "if-then" statement is true, even when the first part is false.

We can break this down into four cases to see this more clearly.

1. The Yankees do win, and they do make the playoffs. The speaker is correct, and the statement is true.

2. The Yankees do win, but they don't make the playoffs. The speaker is wrong, and the statement is false.

3. The Yankees don't win, but they make the playoffs anyway. The speaker is still correct, and the statement is true.

4. The Yankees don't win, and they don't make the playoffs. By convention the statement is still true, because we can't demonstrate its falsity.

Thus, the only way for an "if-then" statement to be false is for the first part to be true and the second part false.

The last connective is a **double implication**, sometimes written "if and only if." This is useful, almost stereotypical, for definitions. "A person is a bachelor if and only if he is an unmarried male." "A number is a perfect square if and only if there is another number that, squared, gives you the first number." "A faculty member can vote on this motion if and only if she is a tenured member of the faculty." Some propositions with this connective have a double implication because they combine two separate implications. For example, "The Yankees will make the playoffs if and only if they win two more games" suggests two things. In this case, one is saying not only that "if the Yankees win two, they will make the playoffs," but also that the only way the Yankees will make the playoffs is by winning two more games: "if the Yankees make the playoffs, they will (have won) two games."

Using these basic semantic primitives, it becomes possible to construct more complicated propositions. "All of the seven dwarfs have a beard except for Dopey," for example, can be built with six conjunctions and one negation:

- Grumpy had a beard
  and
- Sneezy had a beard
  and
- Sleepy had a beard
  and
- Happy had a beard
  and

- Doc had a beard
  and
- Bashful had a beard
  and
- Dopey did not have a beard

It is also possible to devise other connectives in terms of the ideas expressed above. Some logicians, for example, like to distinguish between **inclusive or** and **exclusive or**. These two concepts can be understood as the difference implicit in the following questions:

- Would you like milk or sugar in your coffee?
- Would you like beef or chicken at the banquet?

It is acceptable and even expected that one might answer the first question with "Both, please," but not the second. In the first case, the answer "yes" includes the possibility of both. Similarly, a statement with an inclusive or is true "if and only if" the first disjunct is true, the second disjunct is true, or both are. This is the standard semantics of disjunction discussed above (no one would expect to be served two main courses). Similarly, a proposition with an exclusive or is true if and only if one or the other disjunct is true, but not both. The statement "Timothy is a chinchilla or Sasha is a cat" might therefore be either true or false, depending upon whether "or" is interpreted as inclusive or exclusive—again, we see the focus in mathematical expression on clear definitions to avoid confusion and ambiguity.

The aim, though, is not merely clarity but generality. As with syllogistic logic, we'd like to understand the general structure that governs the truth and falsehood of propositions. To achieve that, we introduce two syntactic changes. The first merely replaces the elementary propositions with **propositional variables** (by convention, capital letters). Thus:

P = "Percy is a cat"

Q = "Quail are a kind of lizard"

R = "Rabelais was the author of both *Gargantua* and *Pantagruel*"

S = "Sasha is a cat"

T = "Timothy is a chinchilla"

The second syntactic change replaces the connectives with symbols:

- $\neg$ means "not"
- $\wedge$ means "and"
- $\vee$ means "or" (and specifically, inclusive or)
- $\rightarrow$ means "if ... then" or "implies"
- $\leftrightarrow$ means "if and only if"[7]

[7]The notation we are describing is the most common, but there is no universal standard. Different writers will sometimes use different symbols to express the

The former change is mostly a matter of concision; if the elementary statements are long, it is simply more convenient to replace them with single letters. The rationale of the latter, however, arises because of the complexities of natural language. Earlier, we had negated phrases like "Percy is a cat" with "It is not the case that Percy is a cat." The reason for such stilted phrasing becomes clearer when these statements are restated in more normative English:

- "Percy isn't a cat"
- "Quail are not a kind of lizard"
- "Rabelais wasn't the author of both *Gargantua* and *Pantagruel*"
- "Sasha isn't a cat"
- "Timothy is not a chinchilla"

As natural as such transformations seem to a native speaker, it is actually rather difficult to define clear linguistic rules about how to convert a sentence into its negation. The negation of "No one entered the room" isn't "No one didn't enter the room," but "Someone entered the room."

---

same logical concept, either for intuitive clarity, or for more prosaic reasons (like the fact that the ¬ key doesn't appear on a standard keyboard). In many programming languages, the symbol "!" is used for ¬, the symbol "&&" for ∧, and the symbol "||" for ∨.

Armed with this syntactic system, we can begin to extend and connect propositions into **formulas**. In propositional logic, a **well-formed formula** (abbreviated **wff**, and pronounced "woof") obeys the following rules:

1. Any propositional variable by itself is a wff.
2. Any wff preceded by the symbol $\neg$ is a wff.
3. Any two wffs separated by the symbol $\wedge$ is a wff.
4. Any two wffs separated by the symbol $\vee$ is a wff.
5. Any two wffs separated by the symbol $\rightarrow$ is a wff.
6. Any two wffs separated by the symbol $\leftrightarrow$ is a wff.
7. Anything that can't be constructed according to these rules is not a wff.

Well-formedness is something like the "grammar" of propositional logic. A grammatical sentence might be false ("Trout live in trees"); an ungrammatical sentence ("In live trees trout") is just gibberish.

All the following are wffs:

- *P* (by rule 1)
  This just represents the original proposition "Percy is a cat."

- *Q* (by rule 1)
  Similarly, this just represents"Quail are a kind of lizard"

- $\neg Q$ (by rule 2 from the previous)
  This represents "It is not the case that quail are a kind of lizard," or equivalently "Quail are not a kind of lizard"

- $P \rightarrow \neg Q$ (by rule 5, from the first and third examples)
  This represents "If Percy is a cat, then quail are not a kind of lizard."

- $\neg Q \vee (P \rightarrow \neg Q)$ (by rule 4, from the third and fifth examples)
  This represents "Either quail are not a kind of lizard or else if Percy is a cat, then quail are not a kind of lizard."

The use of parentheses, which might be added to the list of well-formedness rules, allows us to group expressions into logical units in an unambiguous way. The formula $A \wedge B \vee C$ (A "and" B "or" C), for example, is ambiguous; it could mean either of two things:

- $A \wedge (B \vee C)$: you get A, plus your choice of B or C. (Steak, with either baked potato or fries).
- $(A \wedge B) \vee C$: you have your choice of either a combination of A and B, or you can have C. (You can have the burger and fries, or you can have the salad bar).

Having reduced propositions and their connectives to symbols, there's nothing to prevent us from doing the same thing to the formulas in which they're embedded. By convention, formulas are represented using lower-case Greek letters. Just as the Roman letters P or Q might stand for some longer elementary proposition, so ϕ or ψ might stand for a more complicated formula (e.g. $\neg\neg[(P \vee Q) \rightarrow \neg R]$). This notation allows us to rewrite the syntactic rules above more tersely as:

1. Any propositional variable by itself is a wff.
2. If ϕ is a wff, so is $\neg\phi$.
3. If ϕ and ψ are wffs, so is $\phi \wedge \psi$.
4. If ϕ and ψ are wffs, so is $\phi \vee \psi$.
5. If ϕ and ψ are wffs, so is $\phi \rightarrow \psi$.
6. If ϕ and ψ are wffs, so is $\phi \leftrightarrow \psi$.
7. If ϕ is a wff, so is (ϕ).

It should be noted that there is no limit to the number of different well-formed formulas that can be created, since you can always add another symbol to the front or combine it with another formula.

Truth Tables

The goal, of course, is still to assess the truth or falsehood of propositions. Propositional logic allows us to do this by breaking complex, well-formed formulas into their constituent parts and assessing the consequences of each elementary statement for the truth or falsehood of the overall formula. One of the simplest methods for doing

Table 1.1: Semantics of $\neg$ (not)

if ϕ is	...then $\neg\phi$ is
true	false
false	true

Table 1.2: Semantics of $\wedge$, $\vee$, $\rightarrow$, and $\leftrightarrow$

ϕ	ψ	$\phi \wedge \psi$	$\phi \vee \psi$	$\phi \rightarrow \psi$	$\phi \leftrightarrow \psi$
true	true	true	true	true	true
false	true	false	true	true	false
true	false	false	true	false	false
false	false	false	false	true	true

this involves the construction of a **truth table.** This is simply a representation, in tabular form, of all the possible truth values of the underlying propositions, with the implications for the formulas of interest. Since each proposition can only be true or false, the possibilities are often not difficult to list. (In fact, we did exactly that, although not in tabular format, in our discussion of the semantics of "if-then.")

To begin with a very simple truth table, consider the truth value of the expression $\neg\phi$ as laid out in table 1.1. Right away, we can see that ϕ can only have two possible values, and that each of those two values will result in a particular value when negated. We can therefore "read off" the answer by looking up the initial conditions. The rest of the connectives can be similarly represented, as in table 1.2.

The real power of this representation becomes clear when we use it to map out the possible values of a syllogism. Here is an argument we presented earlier as a weakness of syllogistic logic. Slightly modified:

$$\text{Either Timothy or Sasha is a cat.} \tag{1.34}$$
$$\text{Timothy is not a cat.} \tag{1.35}$$
$$\therefore \text{Sasha is a cat.} \tag{1.36}$$

As a way of formalizing this argument, we note that statement 1.34 is equivalent to the statement "Timothy is a cat or Sasha is cat." We can then represent the statement "Timothy is a cat" by the letter T, and "Sasha is a cat" by the letter S. The formal version of the argument becomes:

$$T \vee S \tag{1.37}$$
$$\neg T \tag{1.38}$$
$$\therefore S \tag{1.39}$$

There are now two underlying propositions (T and S), two premises defined in terms of these prepositions, and one conclusion (which is actually one of the underlying premises). We can now construct a truth table with these five columns:

S	T	T ∨ S	$\neg T$	S
true	true	true	false	true
false	true	true	false	false
TRUE	FALSE	TRUE	TRUE	TRUE
false	false	false	true	false

Recall our earlier discussion of validity. An argument is valid if and only if it is not possible (in any possible world) for the premises to be true and the conclusion false at the same time. This truth table lists all the possible states of the world—S is either true or false, and so is T (independently). We can thus confirm by inspecting the table above that both premises are true in only one of the four possible worlds (the one marked out in capital letters). In this same line, the conclusion is also shown be true. Therefore, in any possible world where the premises are true, so is the conclusion. The argument is demonstrated, using a rote method, to be valid.

While truth tables can be used to test any argument in propositional logic, they can become too cumbersome in some cases. A truth table that involves only a single variable can be represented with only two lines (as in table 1.1 or table 1.3). An only slightly more complex table involving two variables (as above) needs four lines to cover the four possible cases. However, a wff with only three prepositional variables would need, not four or six, but eight lines (four for the case where the third variable was true, four for the case where it was false), and in general, adding another variable will double the number of rows in the table. An argument that began with "One of the seven dwarfs (Grumpy or Sneezy or Sleepy or Doc, etc.) doesn't have a beard" would take more than a hundred lines. An argument that began "one of the 191 UN member states..." would take more lines than there are protons in the universe.

Table 1.3: Equivalence relationships

ϕ	$\neg\phi$	$\neg\neg\phi$	$\phi \wedge \phi$	$\phi \vee \phi$
true	false	true	true	true
false	true	false	false	false

Propositional Calculus

One way to make the system more tractable involves employing a system of inference rules—a method we have already encountered in our discussion of syllogistic logic. The idea, again, is to create a set of rules that can be generically applied to arguments with a particular structure. The **propositional calculus** describes a method of doing this with the particular goal of declaring a broad number of propositions to be formally equivalent. This not only reduces the overall complexity of the system, but makes the reduction itself a matter of rote substitution.

Two statements are said to be **equivalent** if they have the same truth value in all circumstances. For example, the formula ϕ is equivalent to the formula $\neg\neg\phi$, and indeed to the formulas $\phi \wedge \phi$ and $\phi \vee \phi$, as can be seen from the truth table in Table 1.3.

Having done this, we have also shown that the concepts of ϕ and of $\neg\neg\phi$ have exactly the same (logical) meaning. In practical terms, this means that any time one encounters the (sub)formula $\neg\neg\phi$, one can replace it with the simpler ϕ (and vice versa, of course).

So a statement like

$$(\neg\neg(P \vee Q)) \to R \tag{1.40}$$

$$\tag{1.41}$$

can be simplified by replacing $\neg\neg(P \vee Q)$ with its equivalent $P \vee Q$, giving us

$$(P \vee Q) \to R \tag{1.42}$$

$$\tag{1.43}$$

The usual symbol for this replacement is $\vdash$. This observation can thus be formalized as the following two inference rules:

$$\phi \vdash \neg\neg\phi \tag{1.44}$$

$$\neg\neg\phi \vdash \phi \tag{1.45}$$

We can extend this notion slightly to include the idea that if B is true whenever A is true, A $\vdash$ B, even if A and B aren't actually equivalent. For example, we know that if P is true, P $\vee$ Q must be true no matter what Q entails, and therefore P $\vdash$ P $\vee$ Q.

We can also use inference rules like these to define connectives in terms of other connectives, giving us a simpler basis for the analysis of the logic scheme itself. The $\leftrightarrow$ connective, for example, can be shown to be equivalent to a pair of directed implications ($\to$) as in Table 1.4. Finally, and perhaps most importantly, we can show that specific (structural) inferences are valid.

Table 1.4: Redefinition of $\leftrightarrow$

ϕ	ψ	$\phi \leftrightarrow \psi$	$\phi \rightarrow \psi$	$\psi \rightarrow \phi$	$(\phi \rightarrow \psi) \wedge (\psi \rightarrow \phi)$
true	true	true	true	true	true
false	true	false	false	true	false
true	false	false	true	false	false
false	false	true	true	true	true

A particular example of this last kind of inference rule is **modus ponens**. In plain English, this is the structure of such arguments as:

If this is Monday, I have a night class to teach (1.46)

This is Monday (1.47)

∴ I have a night class to teach (1.48)

Now that we know the notation, we can simply say:

$$P \rightarrow Q \tag{1.49}$$

$$P \tag{1.50}$$

$$\therefore Q \tag{1.51}$$

or, more tersely, $[(P \rightarrow Q) \wedge P] \vdash Q$. (If you have the expression "P implies Q" and you have P, you can replace the whole thing with Q.)

Here are some other examples of commonly used inference rules:

- **(modus tollens)** $(P \rightarrow Q) \wedge \neg Q \vdash (\neg P)$

- **(hypothetical syllogism)** $(P \to Q) \land (Q \to R) \vdash P \to R$
- **(disjunctive syllogism)** $(P \lor Q) \land \neg Q \vdash P$
- **(and-simplification)** $(P \land Q) \vdash P$
- **(De Morgan's theorem)** $\neg(P \lor Q) \vdash (\neg P \land \neg Q)$ and $\neg(P \land Q) \vdash (\neg P \lor \neg Q)$
- **(commutivity)** $P \lor Q \vdash Q \lor P$ and $P \land Q \vdash Q \land P$
- **(contrapositive)** $P \to Q \vdash \neg Q \to \neg P$

We could, of course, provide detailed justifications for why these are valid (perhaps using truth tables). In the interest of brevity, we'll leave that as an exercise for the reader. But we can't pass so quickly through this list without noting that the development of these inference rules stands among the more notable achievements of mathematical logic in the last two hundred years or so. They are also very commonly used in mathematical arguments. We won't belabor the point, but it might be useful to meditate on these insights in greater detail once we've finished our overview.

It is possible to define inference rules in terms of other inference rules, and indeed, to prove the validity of any particular inference rule using one or more of the others. Mathematicians have even managed to demonstrate that only one inference rule is actually necessary (modus ponens), and that all others can be derived from it. As elegant as this insight may be, such a restricted rule set can make arguments and proofs extremely confusing. Even so, mathematicians sometimes differ over the number of inference rules they will accept in a specific formulation of propositional logic.

Propositional logic has proven itself to be a powerful tool for thinking about separate propositions. Unfortunately, it suffers from the same sort of limitations that syllogistic logic does—there are obviously valid arguments that cannot be proven in propositional logic (including, amazingly, many of the syllogisms we studied earlier). In particular, propositional logic lacks the machinery necessary to peer inside a simple statement, such as a generalization, and see how the generalization applies in specific instances. Where syllogistic logic could not deal with "or," propositional logic cannot deal with "all" or "some."

1.2.3 Predicate Logic

Predicate logic is a hugely influential system designed to remedy this. It introduces the notion of an expression (called a **predicate**) that expresses an incomplete concept (e.g. "...is a cat" or "...is blue"). It then puts forth a semantics for reasoning about concepts that may apply to some part of the world of discourse but not others. (The term "predicate" is borrowed from grammar, where it refers to the part of the sentence that isn't the subject.)

The logics we've dealt with so far make it easy to deal with complete statements like "Bart Simpson is left handed" (which happens to be true). Predicate logic introduces incomplete statements in order to allow us to work with statements like "x is left-handed" where x is an unspecified variable. The advantages of this become clearer when we try to discuss a particular set of logical statements in which the properties of a set of particular values is in question.

Suppose, for example, that we want to discuss—in an extremely formal and stilted way—the handedness of the members of The Beatles. We could begin by assigning each member of the band to a **constant** (that is, to a symbol, the value of which we do not intend to change). Thus a, b, c and d become John, Paul, George and Ringo, respectively. We'll also replace the partial statement "...is left-handed" with a **predicate variable**: $L(x)$. In this latter formation, we refer to x as the **argument** or **parameter** of L.

Armed with this system, we can now assert that $L(b)$ and $L(d)$, which is simply a more concise way of saying that "b (Paul) is left-handed," and "d (Ringo) is left-handed." Similarly, we can assert $\neg L(a)$ and $\neg L(c)$ (i.e. that neither John nor George is left-handed). And, of course, the connectives have their usual semantics: $[L(d) \vee L(a)] \wedge L(b)$ means that "Ringo or John is left-handed, and Paul certainly is."

Predicate variables need not take only a single argument. Some predicates are inherently relational or comparative, as when we assert that "x is older than y." We could formalize this as the two-place predicate variable $O(x, y)$, and could use this to express ideas like:

$$O(a, b) \text{ (John is older than Paul)} \tag{1.52}$$

which is true, or

$$O(c, d) \text{ (George is older than Ringo)} \tag{1.53}$$

which is false.

We can now introduce two new symbols—called **quantifiers**—that only have meaning in association with incomplete concepts (that is, with predicate variables). $\forall$ means "all" or "every," as in syllogistic logic, and $\exists$ means "some." More explicitly, the expression $\forall x : \phi$ is interpreted as "for all/every/each possible value of x, ϕ is true." $\exists x : \phi$ means that "for at least one value of x, ϕ is true" So of the two statements following

$$\exists x : L(x) \tag{1.54}$$

$$\forall x : L(x) \tag{1.55}$$

statement 1.54 ("Someone [some Beatle] is left-handed") is true, because for at least one value of x—either b or d—the expression is true, but statement 1.55 ("Everyone [every Beatle] is left-handed") is false, because not all the Beatles are left-handed; only some of them are. Similarly, the idea that "no Beatle is older than himself," which is certainly true, can be written as:

$$\forall x : \neg O(x,x) \tag{1.56}$$

We can also express the idea that Ringo is the oldest Beatle ("there does not exist a Beatle who is older than Ringo"):

$$\neg\exists x : O(x,d) \tag{1.57}$$

We can even express the earth-shattering idea that there is an oldest Beatle:

$$\exists y : [\neg \exists x : O(x,y)] \tag{1.58}$$

("There is an unnamed individual y such that there does not exist any individual Beatle x who is older than him.")

Adding a quantifier does not resolve the variable inside the predicate variable (the x in $L(x)$ is still unknown), but it does have an effect on the way we understand that variable. We can therefore distinguish, technically speaking, between two kinds of variables. A **free variable** is a variable that represents an incomplete and underdefined thought, like the x in "x is left-handed." But one can complete the thought by **binding** the variable with something like "for all" or "for some." Thus we would say that the x in "for some x, x is left-handed" is not a free variable, but a **bound variable**.

We can pull all of this into a set of rules for what constitutes a well-formed formula in predicate logic. It is, in essence, the same set of rules we established for propositional logic, but slightly enhanced:

- Any constant, variable, propositional variable, or predicate variable is a wff. (And any variable in this context is a free variable).
- If ϕ is a wff, so are (ϕ) and $\neg\phi$.
- If ϕ and ψ are wffs, so are $\phi \wedge \psi$, $\phi \vee \psi(\phi)$, $\phi \rightarrow \psi$, and $\phi \leftrightarrow \psi$.

- Nothing else is a wff (except as noted below).

To these, we can add the following two rules:

- if ϕ is a wff, and α is a free variable in ϕ, then $\forall\alpha : \phi$ is a wff (and α is no longer free, but bound).
- if ϕ is a wff and α is a free variable in ϕ, then $\exists\alpha : \phi$ is a wff (and α is no longer free, but bound).

Messy, but it serves to capture our intuition of what "all" and "some" really mean.

Nonetheless there is a lot of potential confusion over the exact interpretation of quantifiers, sometimes in areas where natural language is itself ambiguous. For example, these two statements have different truth values in our Beatles example:

$$\neg\exists x : L(x) \tag{1.59}$$
$$\exists x : \neg L(x) \tag{1.60}$$

Formula 1.59 asserts that "it is not the case that a left-handed Beatle exists," or equivalently, that no Beatle is left-handed. This is obviously false since Paul does exist (coded messages on album covers notwithstanding). On the other hand, formula 1.60 asserts that "there exists a Beatle who is not left-handed," which is true. As another example, formulas 1.61 and 1.62 are equivalent (and both false), since they both assert the same thing (that no Beatle is left-handed):

$$\neg\exists x : L(x) \tag{1.61}$$
$$\forall x : \neg L(x) \tag{1.62}$$

(Equation 1.61 states that no left-handed Beatle exists, the second states that all Beatles are not left-handed.)

A more significant issue is that of ambiguity in how quantifiers interact. Using the predicate variable $A(x,y)$ to signify that "x admires y," how would the sentence "everyone admires someone" be written? Perhaps surprisingly, this sentence (in English) has two entirely separate and distinct meanings. The first meaning is that that there is some lucky individual who is universally admired—a global hero or heroine (a statement that is probably not true.) The second is that every person has their own individual person whom they admire, even if that hero is different for every individual. In formal logic, these two interpretations would be written respectively as

$$\exists x : \forall y : A(y,x) \tag{1.63}$$
$$\forall y : \exists x : A(y,x) \tag{1.64}$$

or, in natural language, "there is some individual who is admired by everyone" versus "everyone has an individual whom they admire." It is actually not possible to preserve the ambiguity within the framework of standard predicate logic—a feature that is seen by most mathematicians as a strength, since it reduces confusion and misunderstanding.

Predicate logic has the additional advantage of subsuming both propositional and syllogistic logic. It is, in fact, very easy to represent traditional Aristotelian syllogisms in this framework. The four types of Aristotelian statements can be written as follows (we use the formalization $X(x)$ and $Y(x)$ to represent groups X and Y, respectively):

Type	Example	... becomes
A:	All X are Y	$\forall x : X(x) \rightarrow Y(x)$
E:	No X are Y	$\forall x : X(x) \rightarrow \neg Y(x)$
I:	Some X are Y	$\exists x : X(x) \wedge Y(x)$
O:	Some X are not-Y	$\exists x : X(x) \wedge \neg Y(x)$

We can thus use the truth-generating machinery of predicate logic (which incorporates the machinery of propositional logic) to analyze syllogistic statements.

We could also easily set forth the rules of **predicate calculus.** All the rules we've already seen for propositional calculus (modus ponens, modus tollens, etc.), plus any valid argument in propositional calculus, would remain valid in the extensions. The difference is exactly in the extensions; as the syntax and semantics are extended to incorporate quantifiers, new rules of inference are added to deal with them.

1.2.4 Possible Other Logics

Predicate logic, of course, has its own limitations (the reader is undoubtedly detecting a pattern here). The underlying constants—the object of discussion—can be generalized or quantified, but not the *predicates themselves.* Predicate logic is entirely unable to represent, for example, the statement, "Anything you can do, I can do better" or "I'm not the best in the world at anything" ("For any property P, there is a person x who is better at P than I am"). A logical system that allows quantification of variables but not predicates is referred to as a **first-order logic. Second-order logic** (or more generally **higher-order logic**) extends the system to allow quantification of predicates about variables. Third-order logic permits quantification of predicates about predicates about variables, and so forth. So the statement "For every property P, it's true for a," (or more loosely, "a can do anything."):

$$\forall P : P(a) \tag{1.65}$$

is syntactically and semantically legal only in second-order logic or higher.

Second order logic is useful when you want to reason about not just things, but properties of things. For example, Gottfried Wilhelm Leibniz (1646–1716) put forth (as one of his famous metaphysical principles), the so-called "identity of indiscernibles," which states that two objects are identical if they have all of their properties in common. We can express this idea in second-order logic (but not first-order logic) as:

$$\forall x : \forall y : [x = y \leftrightarrow \forall P : (P(x) \leftrightarrow \neg P(y))] \tag{1.66}$$

Similarly, we can express the related idea that if x and y are different objects, then x has a property that y doesnt:

$$\forall x : \forall y : [x \neq y \leftrightarrow \exists P : (P(x) \wedge P(y))] \qquad (1.67)$$

The difference is subtle, but powerful : in first-order logic, one can quantify over objects (x, y) but not over predicates (P).

Modal logic is an extension of logic to cover the distinctions between certainty and contingency. Experimental results are usually contingent truths (the world might easily have come out some other way), while rules of inference are necessary truths. In modal logic, the symbol $\Diamond\phi$ is used to represent the idea that ϕ is possible, while $\Box\phi$ represents the idea that ϕ is necessary. They are typically related by the following equivalence and rule of inference:

$$\Diamond\phi \leftrightarrow \neg\Box\neg\phi \qquad (1.68)$$

or, informally, ϕ is possibly true if and only if it's not necessarily untrue. That is, "It's possible that Jones is the murderer, if and only if it's not necessary that he's innocent," or, "If it is not possible that Jones is the murderer, then he is necessarily innocent," or

$$\neg\Diamond\phi \leftrightarrow \Box\neg\phi \qquad (1.69)$$

We can, of course, prove that these two formulations are equivalent using the machinery developed for propositional logic.

The idea of modal logic has been extended to cover formalisms involving time (**temporal logic**), of obligation or morality (**deontic logic**), and of epistemology (**epistemic logic**). These are fascinating systems which we will, in the spirit of humility, leave to specialists. That such systems exist as active fields of study and areas for critical thought, though, indicates the overall vitality of the subject. It is likely that many more systems of logic will be developed as people find and formalize additional areas of inquiry. In all cases, however, the major questions are the same: "What can be expressed?" "What can be proven?" "How can things be proven?" and (perhaps most importantly) "Why should we trust the proofs?" We've spent most of this chapter considering the first three questions. The fourth question is important enough to merit a section of its own.

1.3 The Significance of Logic

1.3.1 Paradoxes and the Logical Program

Grelling-Nelson paradox

Earlier, in our discussion of paradoxes, we noted that a paradox usually indicates that something is deeply wrong—not a misapplied method or a mistake, but a flaw in the assumptions and definitions upon which the entire formulation rests. In the late-nineteenth and early-twentieth centuries, it became obvious to practicing mathematicians that many of the intuitive concepts underlying mathematical practice were actually paradoxical. One example of this is the

Grelling-Nelson paradox, which addresses the relationship between the act of naming and the world.

Kurt Grelling (1886–1942) and Leonard Nelson (1882–1927) observed that, at least in some instances, the very act of defining things can create a paradox—which is problematic, given that definitions represent one of the major ways human beings divide up the world. They defined a new pair of words, "autological" and "heterological," as descriptions of adjectives. An "autological" adjective is one that describes itself. For example, the word "terse" is itself terse, the word "pentasyllabic" is itself pentasyllabic, "unhyphenated" is unhyphenated, and "English" is itself English. An adjective is "heterological" if and only if it is not autological; for example "Japanese," "monosyllabic," and "flammable" are not autological and therefore are heterological. Because this definition is an explicit division, every adjective is either autological or heterological.

But what of the word "heterological" itself? If it is itself autological, then it must describe itself and therefore be heterological. Conversely, if it is heterological, then it does describe itself and is therefore autological. Either way, it must (impossibly) both be and not be heterological, yielding the paradox.

This is obviously a close relative of the Barber paradox discussed earlier. As with the Barber paradox, the solution is to reject an assumption. But in this case, the assumption that would need to be rejected is the assumption that one can simply make up definitions and divide the world. The very act of defining would seem to introduce paradoxes into our reasoning.

Russell's Paradox

Similar paradoxes were being discovered relating to many aspects of (then-current) mathematics. In particular, the theory of "sets" at that time relied on an informal understanding that a **set** was just a collection of objects, and that for any property one might care to define, one implicitly defined the set of all objects with that property. Since we know about the idea of "red," we also know about the set of things that are red. Since we know about even numbers, we know about the set of numbers that are even. This notion was formalized by many turn-of-the-twentieth-century mathematicians, most notably Gottlob Frege (1848–1925).

Unfortunately, this naive notion of a set generates almost exactly the same paradox. Known as **Russell's paradox** (after the discoverer, Bertrand Russell, 1872–1970), it involves defining, not a new word, but a new property: that of "not being a member of itself." For example, the set of all things that are not an elephant would include the authors, the reader, the Eiffel tower... and of course that set itself. Similarly the set of all sets (the set of all things that are sets) is a set, and the set of all mathematical concepts is a mathematical concept. On the other hand, the set of all weekdays isn't itself a weekday.

But what happens when we ask about the set of all sets that are not members of itself? Is this set a member of itself?

The result is a paradox. If the set is a member of itself, then it is not a member of itself. If it is not a member of itself, then it must be a member of itself. As with the earlier paradoxes, it indicated to the mathematicians at the time that there was something fundamentally wrong with our naive formulation of set theory.

But if set theory does not follow our intuitions, what else is likely to break? What about the mathematical results that depend upon the ideas and concepts of set theory? How far does the rot spread? If paradox is a sign that we don't understand what we're doing, then a paradox at the heart of mathematics is a problem. For mathematics to progress as a discipline, it needed to be paradox-free. The solution, it seemed, was to define as much of mathematics as possible in terms of a few simple and fundamental principles that could be proven valid and free from paradox. Proofs could then be put on a sound theoretical basis by justifying them in terms of this minimal and rigorous set of assumptions whose truth was beyond realistic question. Mathematics would thus be "axiomatized" and formally validated using specific, valid, logic schemes.

1.3.2 Soundness, Completeness, and Expressivity

Of course, this project would only work if the logic employed was up to the task. In particular, it needed to be **sound**, **complete**, and **expressive**.

We've discussed *expressivity* at some length; there are some statements that simply can't be made, let alone analyzed, in some logics. A logic system capable of supporting all of mathematics would have to be able to express all of mathematics. At a minimum, it would have to be able to express arithmetic. That's the main reason why syllogisms aren't used in formal mathematics much any more—it's too hard even to express simple arithmetic concepts (like the "story problems" one encounters in primary school).

A logical system is *sound* if and only if every argument provable in the system is also valid. In other words, if it can be proved within the system that something is true, then that something is actually true. A system is similarly *complete* if and only if everything true is provable in the system. Thus a sound system does not allow one to prove things that aren't true, and a complete system will not allow unprovable truths to hide in the corners of the system.

Syllogistic logic provides a good example of this. It can be proven (though we won't, for the sake of brevity) that the "standard" syllogistic logic is both sound and complete. But let's suppose that some mad mathematician wanted to incorporate the following argument as a legitimate structure (we'll call it the **Baloney* syllogism):[8]

$$\text{Major premise–I No X are Y} \tag{1.70}$$
$$\text{Minor premise–I No Y are Z} \tag{1.71}$$
$$\text{Conclusion–I} \therefore \text{No X are Z} \tag{1.72}$$

Obviously, the **Baloney* syllogism is invalid, since it permits nonsensical inferences like "No trout are snails. No snails are fish. Therefore, no trout are fish." We can do worse than that. By letting

[8]In logic, it is customary to mark deliberately spurious arguments with an asterisk in order to indicate the author's awareness of the wayward logic being employed. The practice is also common in linguistics, where it is used to mark grammatically "illegal" sentences used to illustrate a grammatical point. **We will this practice to adhere now from on.*

X and Z represent the same concept, we can prove that nothing is itself—"No dogs are cats. No cats are dogs. Therefore, no dogs are dogs." But beyond that, any system that incorporates the **Baloney* syllogism is unsound, because we don't know whether any specific conclusion drawn from this system is true or not. The system itself permits the "proof" of false statements.

Now, suppose that the hypothetical mathematician goes completely crazy and insists not just that **Baloney* is a legitimate rule of inference, but also that it is the only rule that can be used. This new system is not only unsound but incomplete, because there are traditional arguments that are otherwise known to be valid that can no longer be made. From the following list of observations:

- All A are B
- No B are C

a traditional logician could infer that "No A are C" (cf. Darii), but our madman could not. There are thus not only untruths that are provable (the system is unsound) but truths that are unprovable (the system is incomplete).

Any system that is not complete can (presumably) be enhanced by the addition of new inference rules, and similarly, an unsound system can have its ruleset trimmed to eliminate fallacious conclusions. Combine that with a system sufficiently expressive as to allow one to express any concept, and one presumably has a perfect logical system. The key question, then—and at the turn of the last century, a question of world-wide interest—is whether such a system is possible. At the International Congress of Mathematicians in Paris in

1900, the great German mathematician David Hilbert (1862–1943) famously proposed it as one of twenty-three open, unsolved problems for twentieth-century mathematics.

Thirty-one years later, the mathematician Kurt Gödel (1906–1978) gave an answer that pointed to a deep fundamental truth about the nature of mathematics. It remains one of the most startling results ever proposed by a mathematician, and has been the source of much (occasionally facile) musing among philosophers ever since.

The answer to that key question, you see, was "No."

1.3.3 The Greatest Hits of Kurt Gödel

To understand the full significance of Gödel's theorems, we have to go back to Russell. Russell, motivated in part by his own paradox, undertook to write a colossal work that would set mathematics on a firm foundation once and for all. The machinery he employed was cumbersome and hard to understand, the notation verged on the inscrutable, and the amount of detail was fearsome. It was also slow going; one particularly memorable line stated, "From this proposition it will follow, when arithmetical addition has been defined, that 1+1=2"—a conclusion offered on page 360. Yet for all that, the work seemed entirely sound. It began by formally defining set theory in terms of predicate logic. By the end of the third volume, he and his co-author Alfred North Whitehead (1861–1947) had laid down apparently paradox-free foundations, not only for set theory and logic, but for arithmetic, algebra, and the basics of analysis and calculus. They called it, in an extraordinary moment of hubris, *Principia Mathematica* after Isaac Newton's rather famous work of that name.

But the question remained: were there unseen paradoxes and contradictions in the *Principia?* In particular, were there any statements in the *Principia* that could be proven to be both true and false using the mechanisms of the *Principia* itself? Were there any mathematical statements that *couldn't* be proven to be either true or false?

Gödel's Findings

Gödel was able to address this question directly in a way that shattered the entire program in a series of three very famous and influential statements.

Gödel's **Completeness Theorem** showed that **first-order predicate logic** was complete (it had long been known to be sound). Unfortunately, first-order predicate logic itself was not expressive enough to support the whole of mathematics, and in fact, Gödel's proof of his Completeness Theorem couldn't be expressed in first-order predicate logic. He needed to use something stronger.

Gödel's **First Incompleteness Theorem** showed that any formal system of logic powerful enough to support mathematics—or arithmetic, for that matter—had to be either incomplete or unsound, and specifically inconsistent. He was able to construct a single statement and show that if that statement was provable, so was its opposite. Therefore, either both statements were provable (meaning the system was unsound and inconsistent) or neither were (and the system was incomplete).

Gödel's **Second Incompleteness Theorem** showed that it is not possible for a consistent system powerful enough for arithmetic to prove its own consistency, or alternatively, that any system capable of proving its own consistency is inconsistent (making the proof wrong).

What Gödel Really Found

Obviously, one can't summarize in a few paragraphs or pages all the implications of some of the most profound results in the history of mathematical logic. But some discussion is needed, if for no other reason than to cut through some of the philosophical misunderstandings that have grown up around these theorems.

The Completeness Theorem is actually fairly straightforward. Essentially, Gödel was able to show that first-order logic did really reflect the inherent idea of the possible-world semantics defined earlier, and that if one formalizes the idea of "possible worlds" strictly in terms of constants and (first-order) predicates, any statement true in all possible worlds can be proven using the machinery of first-order logic.

The First Incompleteness Theorem however, shows that if we permit second-order and higher predicates, it is possible to generate paradoxes within the system. The paradox that Gödel was able to generate is a variation on a very old paradox called the **Liar Paradox**.

The Liar Paradox is attributed to (the possibly mythical) Epimenides of Crete (6th c. B.C.E), who supposedly stated that "All Cretans are liars." Of course, if this statement were true, then Epimenides himself would have to be a liar, and thus this statement would be a lie, which means that the statement can't be true. There are a few linguistic holes in this argument, but they can be patched by restating the paradoxical sentence in this form:

- This sentence is false.

If the sentence is true, then of course, it must be false. On the other hand, if the sentence is false, then it must be the case that the sentence is not false, i.e., true. Either way, the assumption that the sentence is true or false yields a contradiction, so the sentence is paradoxical.

Gödel was able to construct a similar proposition that stated:

- This sentence is not provable (in this logical system)

Obviously, if the sentence can be proved, then it must be true. But if it is true, then it can't be proved. So if the statement both can and cannot be proved, we have our paradox. Similarly, if the sentence can't be proved, then it is true but unprovable—but in this case we have an unprovable truth.

The brilliance of his proof lies not merely in identifying this as a potential paradox, but by finding a way to frame this sentence purely in terms of arithmetic properties. His constructed sentence is really just a huge formula that is either true or false, but that never actually says anything about "this sentence" or "provable." It's therefore possible to construct such a sentence in any logical system powerful enough to express simple arithmetic ideas like numbers, addition, multiplication, and so forth.

The implications of this are profound. In essence, any system of logic one proposes must fall into one of three categories. It may be too weak and inexpressive to be able to handle primary school mathematics (which makes it useless for the logicization program defined above). It may be unsound, in which case we cannot rely on the proofs. Or it may simply be incomplete, in which case it is still a usable system, but there are truths in mathematics that are inaccessible to it. The Second Incompleteness Theorem takes this further, and shows that any system that can prove its own consistency must be asserting an incorrect proof, and therefore must fall into the second category above.

This finding, as one might expect, dashed the hopes of many mathematicians who had expected to find a perfect logical system. Instead, they found that the best they could make was a conservatively imperfect logical system, one that could find and validate some truths, but not all of them. Later researchers were even able to find specific examples of statements that could not be proven true or false within the framework of the *Principia.* On the other hand, the very fact that *Principia* is incomplete offers a certain odd assurance. If it were, in fact, unsound, it would be possible to prove anything at all. So the fact that there are a few unprovable statements lends a sort of contrastive credence to what we *can* prove.

These results were hugely influential within the mathematics community, as one might expect. Gödel himself is often considered to be one of the greatest mathematicians of the twentieth century (if not of all time). But precisely because of the revolutionary nature of these findings, his theorems have been enlisted (at times, by otherwise reputable thinkers) in the project of making broad claims about the limitations of science, language, computers, and the human mind. Much of this is fatuous in large part because Gödel's theorems have mostly narrow practical implications and are concerned exclusively with the limitations of mathematical logic.

But even if the implications of Gödel's theorem were, within a broad philosophical perspective, somewhat narrow, the larger cultural implications remain. Hilbert had announced his "unsolved problems" with all the faith and high-minded exuberance of the nineteenth century:

> Who among us would not be happy to lift the veil behind which is hidden the future; to gaze at the coming developments of our science and at the secrets of its development in the centuries to come? What will be the ends toward which the spirit of future generations of mathematicians will tend? What methods, what new facts will the new century reveal in the vast and rich field of mathematical thought? [6]

Within a few decades, the centuries-old humanistic program that had begun with the Enlightenment (itself an era of profound mathematical insight and progress) would give way to a far more skeptical vision of human possibility. If Newton's calculus had metaphorically typified the Age of Reason, so Gödel's theorem would come to represent the problematic aspirations of the Atomic Age. It wasn't that logic had its limitations (not even the most naive medieval scholastic had doubted that). It was that for the mightiest of all reasoning systems—mathematics—it could never be otherwise.

2 Discrete Mathematics

The Hungarian mathematician Paul Erdös (1913–1996) spent most of his adult life traveling from one university to another, working on mathematical problems with friends and colleagues, and living out of a suitcase that contained nearly all of his worldly possessions. One of mathematics's most legendary eccentrics (and that is saying something), Erdös has the distinction of being the most prolific mathematician in history—the author of over fifteen-hundred papers in areas that range over most of the subjects in this book.

As a tribute to Erdös, the custom emerged of assigning "Erdös numbers" to research mathematicians. A mathematician who co-authored a paper directly with Erdös (an honor held by over five hundred people) is assigned an Erdös number of 1. People who have collaborated with a member of this group (but not with Erdös himself) have an Erdös number of 2. If you collaborated with someone who collaborated with someone who has a 1, you're a 3 (and so on).

Having a low Erdös number makes for good bragging rights, and some have suggested that such proximity to the well known may serve as a reasonable indicator of importance within a given field.[1] What is most fascinating, however, is the fact that the average Erdös

[1] You'll be relieved to know that one of the authors of this book is a 4.

number is less than 5. This suggests that the network formed by the group who have Erdös numbers exhibits what is sometimes called the "small world" phenomenon. The network itself is quite large, and most of the people in the network are not neighbors. Yet it only takes a few hops to get from one node to another. The game known as "Six Degrees of Kevin Bacon"—a variant of the Erdös-number concept in which one tries to form connections between actors who have appeared in a movie with Kevin Bacon—also demonstrates the concept. As with Erdös numbers, the strange thing is how easy it is to find a connection; even moderately knowledgable movie buffs can "win" the game with relative ease.

Speaking of small worlds: It was one of Erdös's countrymen, the Hungarian fiction writer Frigyes Karinthy (1887–1938), who appears to have been the first to suggest that everyone in the entire world is connected by "six degrees of separation." He put this idea forth in a 1929 short story entitled "Chains," in which people at a party play a game that is remarkably like Six Degrees of Kevin Bacon.[2] Frigyes's story is apropos in more ways than one. The period between the wars witnessed a marked popular interest in optimization, statistics, and demographics (Taylorism is at its zenith and the term "robot" is coined during this period). There was also a widespread sense, evident both in art and in public discourse, that the world was getting smaller. Frigyes's game managed to encapsulate both spirits of the age, and perhaps to anticipate a similar

[2]The phrase "six degrees of separation" became the proverbial way of describing the phenomena as a result of John Guare's 1990 play of the same title. The characters in Frigyes's story actually set the number at 5.

concern with such matters in our own day. The book in which the story appears is fittingly entitled, *Everything is Different*.

In this chapter, we continue a tradition in mathematics of grouping together a broad set of loosely related subjects under the general title "discrete mathematics." According to the usual line of reasoning, discrete mathematics is "discrete" because it deals with structures that are separate, distinct, and enumerable (like the whole numbers, the objects in a set, or the nodes on a network). This is as opposed to continuous mathematics, which traffics in things that are, for lack of a better term, "smooth" and "gapless" (like the real numbers, the path of a rocket, or the curve of a butterfly's wing).

This is a useful division, as far as it goes—a way to separate the (only apparently) disparate realms of, say, set theory and calculus—but it may be more useful (and in the end, no less rigorous) to think of discrete mathematics as a group of topics that have the same feel to them. Discrete mathematics is the branch of mathematics most requisite for understanding the deep mysteries of computation, including such rarefied matters as microchip architecture, data structures, algorithms, networks, and cryptography. Yet it is also the world's leading source of games and puzzles (and therefore, an obligatory subject for students of games and game mechanics). Problems in discrete mathematics often come down to gaining some insight into the nature of a system. And when you have that insight, it can feel like you just figured out a trick or discovered a secret. People work out their Erdös numbers and determine how close an actress is to Kevin Bacon because it's fun. Yet it is also possible to ask whether everyone in the world really *is* separated by only a few nodes in a vast network—a property of the social life of human beings that may have quite extraordinary implications. In either case, the object of study involves things that are made of discrete, enumerable elements.

2.1 Sets and Set Theory

2.1.1 Set Theory, Informally

We begin, then, with a discussion of discrete things so fundamental that it has very often been thought of as the foundational concept of mathematics itself—namely, the **set**, which we can informally define as "a collection of objects."

The easiest way to define any specific set is to list the objects that are in it (usually between braces: $\{\}$). The following, for example, is a set of (the names of) musicians:

$$\{\text{John}, \text{Paul}, \text{George}, \text{Ringo}\}$$

Here's one that contains musical instruments:

$$\{\text{guitar}, \text{bass}, \text{drums}, \text{kazoo}\}$$

We can have a set of letters:

$$\{a, e, i, o, u\}$$

...or a set of early Christians:

$$\{\text{Matthew}, \text{Mark}, \text{Luke}, \text{John}, \text{Paul}\}$$

This is not the most convenient notation for large sets; the set of all Parisians would take pages to list, and the set of odd numbers $\{1,3,5,7,9,\ldots\}$ never ends. In such cases, it becomes useful to describe sets in terms of the properties their elements share.[3] For example, instead of:

$$\{1,3,5,7,9\ldots\}$$

we could write:

$$\{x : x \text{ is an odd number}\}$$

(This is sometimes read out loud as "x *where* x is an odd number."). Similarly, we could write:

$$\{x : x \text{ is a vowel}\}$$

instead of

$$\{a,e,i,o,u\}$$

The most important attribute of a set, is that it has **elements** or **members**. The elements of {Matthew, Mark, Luke, John, Paul} are Matthew, Mark, Luke, John, and Paul (as you might expect). On the

[3]As it turns out, it is only possible to describe sets accurately in terms of their properties some of the time. The cases in which it is not possible—which tend to occur only in the context of extremely formal set theory—constitutes the rare moments in which Russell's paradox appears. The set theory we're describing here, which makes liberal use of natural language while at the same time avoiding paradoxical cases, is often called **naive set theory**.

other hand, neither Ringo nor George is a member of that set, and neither is the number 5.

If all the elements of one set are also elements of another, we say that the first set is a **subset** of the second, or alternatively, that the second is a **superset** of the first. The set

$$\{\text{Matthew}, \text{Mark}, \text{Paul}\}$$

for example, is a subset of

$$\{\text{Matthew}, \text{Mark}, \text{Luke}, \text{John}, \text{Paul}\}.$$

The set of all odd numbers ($\{1,3,5,7,9\ldots\}$) is a superset of the smaller set $\{1,3,5,7\}$. Note that under this definition any set is both a subset and a superset of itself. After all, any element in a set (treating it as the first set) is an element in that set (treating it as the second). In this situation, we say that the two sets are **equal**. The set {John, Paul, George, Ringo} is neither a subset nor a superset of {Matthew, Mark, Luke, John, Paul}, and so we say that these two sets are not **comparable**.

So is {a, e, i, o, u} a subset of the set of all vowels? Or to put it another way, is a whole pie "part of" a pie? From a technical standpoint, every morsel of the pie is part of the entire pie, so, mathematically, the answer would be "yes." Intuitively and lingustically, this isn't what we mean by "part of" (or subset). To allow for the intuitive meaning of superset and subset, we speak of a set being a **proper subset** (or **proper superset**) of another if-and-only-if the set is a subset (superset), but *not* equal. So {John, Paul, George, Ringo} is a subset, but not a proper subset, of {John, Paul, George, Ringo}.

2.1.2 Set Theory, Notation

As with logic, there is a set of more-or-less standard symbols that make writing about sets a little easier and faster (if slightly more opaque to the uninitiated). To start with, we can use capital letters to indicate specific sets, as follows:

$$\begin{aligned}
B &= \{\text{John, Paul, George, Ringo}\} && (2.1)\\
I &= \{\text{guitar}, \text{bass}, \text{drums}, \text{kazoo}\} && (2.2)\\
V &= \{a, e, i, o, u\} && (2.3)\\
C &= \{\text{Matthew, Mark, Luke, John, Paul}\} && (2.4)\\
O &= \{1, 3, 5, 7, 9, \ldots\} && (2.5)
\end{aligned}$$

The symbol $\in$ (as in $x \in S$) means "is an element of" ("x is an element of S"). So instead of writing out that "Ringo is an element of the set {John, Paul, George, Ringo}," one can simply write "Ringo $\in B$." The symbol $\notin$ means "is not an element of." So $a \in V$ (since a is a vowel), and $3 \in O$, (since 3 is an odd number) but Ringo $\notin C$ (since there wasn't an apostle named Ringo, as far we know).

The symbol $\subseteq$ means "is a subset of." So Instead of writing "S is a subset of T," we can write "$S \subseteq T$." Logic and set theory are closely related, and the symbols are often combined. We can, for example, take this formal definition:

> "If and only if for every x in a set S, x is also in the set T, then S is a subset of T."

and compact it down to:

$$(\forall x : x \in S \rightarrow x \in T) \leftrightarrow S \subseteq T$$

Unpacking that line, we get the following:

- $\forall x$: "For every x"
- $x \in S \rightarrow x \in T$ "if x is an element of S, then x is an element of T"
- $\leftrightarrow$ "if and only if"
- $S \subseteq T$ "S is a subset of T."

We can also express the idea of a set *not* being a subset of another with the symbol $\not\subseteq$. If we wish to talk about proper subsets, we use a slightly different symbol: $\subset$. Note that the relationship between $\subset$ and $\subseteq$ is the same as that between $<$ and $\leq$. As you might expect, the symbols $\supseteq$ and $\supset$ mean "is a superset of" and "is a proper superset of."

For set equality (which holds when two sets have the same elements), we simply use $=$. So the following sentences are all true (when S and T are sets). Think about them until you feel comfortable with the notation and with the concepts they represent.

- if $S = T$, then $S \subseteq T$
- if $S = T$ and $T = R$, then $S = R$

- if $S \subset T$ then $S \subseteq T$, but not necessarily the other way around.
- if $S \subset T$ and $T \subset R$, then $S \subset R$
- if $S \subset T$ then $T \supset S$
- $S \not\subset S$, always.
- $S \subseteq S$, always.
- if $x \in S$ and $y \in T$, but $x \notin T$ and $y \notin S$, then $S \nsubseteq T$ and $T \nsubseteq S$.

The set that contains no elements is called the **empty set** or (sometimes) the **null set**. Its symbol is $\emptyset$. We use it to denote the set of all female US presidents before 1900, the set of Kurdish-speaking English kings, the set of four-sided triangles, and so forth. By definition, for any x, $x \notin \emptyset$. But oddly enough, for any set S, $\emptyset \subseteq S$. This follows logically (if counterintuitively) from the definition of a subset. We need to assume that every element of $\emptyset$ is in a given set S, because if this were not true, there would have to be at least one element in $\emptyset$. But since there are no elements in $\emptyset$, there are no elements that are not in S, and that's close enough to saying that every element of $\emptyset$ is in S. You are not the first person to think that this is a trying bit of sophistry, but set theory tends to break apart without it.

In addition to these simple symbols, mathematicians have defined a few more to handle simple set operations. The **intersection** (written $\cap$, as in $B \cap C$) of two sets is the set of all elements that they have in common. For example John is the name of both a Beatle and an Evangelist, so sets B and C both contain the name John. This could be written, formally, as "John $\in B \cap C$." The **union** (written $\cup$) is the set of all elements in either one, so the union of sets B and C includes people who are only Beatles (like George), as well as people who are only apostles (like Matthew). So in this notation, and using the sets defined above:

$$
\begin{aligned}
B \cup C &= \{\text{Matthew}, \text{Mark}, \text{Luke}, \text{John}, \text{Paul}, \text{George}, \text{Ringo}\} && (2.6)\\
C \cup B &= B \cup C && (2.7)\\
B \cap C &= C \cap B = \{\text{John, Paul}\} && (2.8)\\
O \cap V &= \emptyset && (2.9)\\
V \cap V &= V \cup V = V = \{a, e, i, o, u\} && (2.10)
\end{aligned}
$$

There is a strong (and deliberate) overlap between these symbols and the ones used in logic; the $\vee$ symbol (meaning "or") deliberately mimicks the traditional $\cup$ symbol (meaning "union") and the $\wedge$ symbol ("and") follows $\cap$ ("intersection"). The connection is fairly simple; an element is in the union of sets $S \cup T$ if-and-only-if it is in set S or ($\vee$) it is in set T (the case for $\cap$ and $\wedge$ is similar).

The symbol $|S|$ is used for the **cardinality** of S, which is just an ornate term for **size**. It refers to the number of elements in S. (I.e. $|\emptyset| = 0$ and $|B| = 4$).

Finally, the expression $S - T$ refers not to ordinary subtraction, but to the related concept of **set difference**. The difference of two sets is simply the elements that are in the first but not the second. Therefore:

$$B - C = \{\text{George, Ringo}\} \tag{2.11}$$
$$C - B = \{\text{Matthew, Mark, Luke}\} \tag{2.12}$$
$$V - V = \emptyset \tag{2.13}$$
$$O - \emptyset = O \tag{2.14}$$

We can also talk about $-T$: the "set of all things that aren't in T," though the use of this notation is not consistent. Some authors use $-T$, while others prefer T', T^C or $\bar{T}$. Conceptually, it doesn't matter much which one you use.

2.1.3 Extensions of Sets

As defined, sets are useful enough to represent many real-world things. There are cases, however, where sets are simply not "structured" enough to describe the properties that inhere in particular collections of objects. The two obvious examples of this involve ordering and multiple elements.

As formally defined, sets have no internal order. This is formalized in standard theory as the **axiom of extensionality:** *if two sets X and Y have the same elements, then X* = *Y*; it doesn't matter whether I put the peanut butter or the jelly onto my sandwich first. But sometimes one needs internal order: the participants in a race, for example, finish in a specific order. To support this, we can introduce the idea of an **ordered pair**—a collection of two elements in a specific order.

For example, the Olympic gold medalist at the 2008 men's 200m butterfly was the astonishing Michael Phelps. The silver and bronze went to Laszlo Cseh and Takeshi Matsuda, respectively. If Cseh had been seven-tenths of a second faster, he would have won gold (and Phelps would have won silver). The set of medalists would have been the same, but of course, the order of finishing would have been different.

The notation for ordered pairs is similar to that used for sets; a set is written with curly braces $\{\}$, while an ordered pair is written with parentheses (). So $\{a,b\} = \{b,a\}$ (since both are sets and the elements are the same), but $(a,b) \neq (b,a)$ (since both are ordered pairs and the order is different). And of course, $\{a,b\} \neq (a,b)$ (one is a set, the other an ordered pair). Since ordered pairs can easily be assembled into a special kind of set, most of the useful properties of sets are also true for ordered pairs. The generalization from ordered pairs to ordered triples, quadruples, and so forth is fairly simple, and mathematicians will often speak of an ordered **tuple** when the exact number of elements is unspecified or unimportant. To continue the example above, the set of medalists {Cseh, Matsuda, Phelps} would have remained the same, but the ordered triple (Phelps, Cseh, Masuda) would have changed to (Cseh, Phelps, Matsuda) if Laszlo Cseh had been that tiny bit faster.

The same item sometimes appears several times in the same set-like collection; the Beatles' instruments, for example, were {guitar, bass, guitar, drums}. Although this is disallowed in a set, it is allowed in a generalization called a **multiset**. This is also easy to create using set primitives; a **multiset** is a set of ordered pairs (a,N), where a is the element and N is the number of times it occurs (the

multiplicity).

2.2 Relations and Functions

One of the most important extensions to the notion of a set is the ability to make sets of sets (i.e. sets whose members are themselves sets). We have, for example, just defined an ordered pair as a kind of set. It is an easy step from there to thinking about sets whose members are themselves ordered pairs—which in turn leads us to a particularly powerful mathematical object called a **function**. Before we get to that, however, we need to mention a few additional bits of machinery.

2.2.1 Cartesian Products

We first define the **Cartesian product** of two sets as the set of all possible ordered pairs in which the first element of the pair is drawn from the first set, and the second element of the pair is drawn from the second set. So if we have a set B that contains the members of The Beatles ($\{\text{John}, \text{Paul}, \text{George}, \text{Ringo}\}$) and a set I that contains instruments ($\{\text{guitar}, \text{bass}, \text{drums}, \text{kazoo}\}$), the set $B \times I$ will be the combination of all possible Beatle-members with all possible instruments:

$$B \times I = \begin{cases} \{(\text{John}, \text{guitar}), (\text{John}, \text{bass}), (\text{John}, \text{drums}), (\text{John}, \text{kazoo}), \\ (\text{Paul}, \text{guitar}), (\text{Paul}, \text{bass}), (\text{Paul}, \text{drums}), (\text{Paul}, \text{kazoo}), \\ (\text{George}, \text{guitar}), (\text{George}, \text{bass}), (\text{George}, \text{drums}), (\text{George}, \text{kazoo}), \\ (\text{Ringo}, \text{guitar}), (\text{Ringo}, \text{bass}), (\text{Ringo}, \text{drums}), (\text{Ringo}, \text{kazoo})\} \end{cases} \tag{2.15}$$

The product $I \times B$ would be almost the same, except that the instrument would come first in each pair (so (John, bass) $\in B \times I$, but (John, base) $\notin I \times B$). Note that if B has x elements (that is, if $|B| = x$) and I has y elements, then $B \times I$ has $x \cdot y$ elements—a fact that will prove of some importance later on.

The formal notation looks like this:

$$A \times B = \{(a,b) : a \in A \wedge b \in B\} \tag{2.16}$$

If you can understand this hieratic bit of symbolic notation, you're ready to proceed.

2.2.2 Relations

A **relation** is a subset of a Cartesian product. If $B \times I$ are all the possible pairings of Beatles with instruments, we can let P be the subset ($\subset B \times I$) of instruments they actually played. Thus, (Paul, bass) $\in P$, but (George, kazoo) $\notin P$. In this framework, we say that Paul is "related to" bass, George and John are both "related to" guitar, and Ringo to the drums. We could write this as Paul$P$bass, but since this is unduly confusing even in the context of mathematical notation, mathematicians will usually not name relations using letters, but instead find some underutilized bit of typography. For example, we might use $a \diamond b$ to represent the idea that a "is related to" b (and therefore, Paul $\diamond$ bass).

We can then formally say that for any relation P,

$$a \diamond b \leftrightarrow (a,b) \in P \tag{2.17}$$

or in other words, that a is related to b if and only if the pair (a,b) is in P.

Relations are a simple and mathematically elegant way to describe the properties of objects as well as relationships among objects. For example, the following are examples of relations:

- if A is a set of items in a store's inventory, and B is a set of prices, $\{(x,y) : y \text{ is the sales price of item } x\}$ is a relation on $A \times B$.
- if A is the set of people, then $\{(x,y) : x \text{ is married to } y\}$ is a relation on $A \times A$ (sometimes abbreviated as just "a relation on A").
- if A is a set of animals in a zoo, and B is a set of foodstuffs, then $\{(x,y) : x \text{ eats } y\}$ is a relation on $A \times B$.
- if A is a set of departure cities and B a set of arrival cities, then $\{(x,y) : \text{There is a direct flight from } x \text{ to } y\}$ is a relation on $A \times B$.

- if, in the example above, sets A and B are the same (every arrival city is also a departure city and vice versa), then $\{(x,y)$: There is a direct flight from x to $y\}$ is a relation on $A \times A$ (or "a relation on A" using the same abbreviation).

This might all seem like a tedious way of stating the obvious. But this exacting, mathematical way of expressing things is at the root of some of the most useful computational tools around. Online flight ticketing systems, for example, use relations like the fourth one above to determine the best way to get from one city to another. If Boston $\diamond$ Pittsburgh ("Boston is related to Pittsburgh"), you can get from here to there in a single flight, but otherwise you will have to change planes. Similarly, we can ask the question "What do elephants eat?" using the language of relations (in formal notation, "what are the x such that (elephant, x) is in the eating-relation?" or "what are the x such that elephant $\diamond$ x?") This is precisely how relational databases work; they determine set membership using "relations."

2.2.3 Functions

We have no reason to assume that elephants eat only one thing, or that only one thing eats insects. There are, however, a large class of things where this relationship does hold: we can probably assume that an object in a store has only one price, and that there is only one combination of numbers that will open a particular combination lock. This latter relationship is an important one—so important, that it gets its own name.

To formalize the concept a little further, we can introduce the idea of domain and range. The **domain** of a relation (R) is the collection of all the first elements of the pairs (in other words, $\{x : (x,y) \in R\}$). The **range** is the collection of all the second elements: ($\{y : (x,y) \in R\}$). The domain of the first example above is the store's entire inventory, while the range is the set of all the prices. (If the store is called "Everything's A Dollar," for example, we expect the range to be the **singleton** set $\{\$1\}$.)

A relation is called a **function** when there is a *single* element of the range related to each domain element. This captures our idea that every item has a price, and each item has only one price. In fact, it captures our intuitive notion of the properties of an item—we can talk about *the* price of an orange or *the* color of a car or *the* address of a house, in each case confident that a given item has only one such attribute related to it. (By contrast, *the* ancestor of a person is not a single set. Father-of is typically a function, but parent-of, brother-of, child-of, or ancestor-of would not be.)

Because functions have only one range element associated with each domain element, there is an alternative notation often used specifically for functions. Instead of writing the relation symbol in between the two elements,[4] we write the relation symbol and the domain element and use that to represent the corresponding range element. So, if Paul $\diamond$ bass and $\diamond$ is a function, then we can write $\diamond$ Paul or $\diamond$ (Paul) to represent his bass. This is also the notation for "standard" mathematical functions like $\cos\theta$ or computer functions

[4]Which we do more often than you might realize. The $\geq$ in $4 \geq 2$ indicates a relation on a set of numbers.

like sqrt(x) (which is typewriter-speak for $\sqrt{x}$).

In fact, this is the notation used for the kind of "function" you may recall from algebra class. When one writes an equation like

$$y = Ax + B$$

or even

$$f(x) = Ax + B$$

one is defining a relation f between x and y, such that x is related to y if and only if $y = Ax + B$. In other words, f is a collection of ordered pairs of numbers, and the pair (x, y) is in the collection f only if the algebraic relation holds. Furthermore, we know that this relation is a function because there is only one y value for any given x.

2.2.4 Equivalence Relations and Partial/Total Orders

Relations have properties that distinguish one from another. Continuing to use the $\diamond$ symbol as a general "is related to," we would say that any relation $\diamond$ is **reflexive** if every x is related to itself.[5] For example, the relation $\{(x,y) : x \text{ has the same name as } y\}$ is reflexive; you obviously have the same name as yourself. On the other hand $\{(x,y) : x \text{ is the father of } y\}$ is obviously not reflexive.

[5]In formal notation: $(\forall x : x \diamond x)$. Remember that $\forall$ means "for every" or "for all."

A relation is **symmetric** if whenever x is related to y, y is related to x and vice versa ($\forall x, y : x \diamond y \leftrightarrow y \diamond x$). "Is married to" is a good example of a symmetric relation, but notice that "is the father of" is not symmetric. Finally, a function is **transitive** if whenever x is related to y, and y is related to z, x is related to z ($\forall x, y, z : x \diamond y \wedge y \diamond z \rightarrow x \diamond z$). Neither "is married to" nor "is the father of" is transitive, but "has the same name as" and "is taller than" are both transitive.

Using this vocabulary, you can see that our intuitive notion of "equality" is reflexive, symmetric, and transitive. We can turn this observation around and say that any relation that is reflexive, symmetric, and transitive behaves like equality. "Has the same name as" behaves like equality in this regard.[6] Such a relation is called an **equivalence relation**, and it can be used as a very powerful analogue of equality. The notion of set equality (two sets are equal if they have the same members) is an example of such an equivalence relationship defined over sets. The most common use of an equivalence relation, however, is to separate (**partition**) the elements of a given set into subsets where all the elements of a given subset are related to each other, but two elements in different subsets are not related. In the context of an equivalence relation, these subsets are called **equivalence classes**.

As an example, let's define a relationship ($\sim$) on the set of numbers (integers) as follows: ($x \sim y$) if and only if the (numeric) difference $x - y$ is an even number. So $1 \sim 3$ and $3 \sim 5$, but $1 \not\sim 6$.

[6]Though a badly programmed computer might not know the difference between three different people named "John Smith"—a matter which has facilitated identity theft in an alarming number of cases.

Inspection (that's math-speak for "take a look, and you'll see.") reveals that $\sim$ is an equivalence relationship—that is, it is reflexive, symmetric, and transitive. In more detail:

- For any number n, $n - n = 0$. Since 0 is even, $n \sim n$. (Reflexive)
- For any numbers n and m, if $m - n = x$, $n - m = -x$. If x is even, so is $-x$. Therefore, if $m \sim n$, then $n \sim m$. (Symmetric)
- For any numbers m, n and o, if $m - n$ is x and $n - o$ is y, $m - o = m - n + n - o = (m - n) + (n - o) = x + y$. If x and y are even, so is $x + y$. Therefore, if $m \sim n$ and $n \sim o$, then $m \sim o$. (Transitive)

What this relation ($\sim$) really does is partition the set of numbers into two equivalence classes: the class of odd numbers and the class of even numbers. All of the odd numbers are related to each other (but not to any even numbers) and vice versa for the even numbers. In general, equivalence relationships allow us to capture the idea of a group of interchangeable objects that are nevertheless distinct from other kinds of items. Money is like this. Any $5 bill is typically interchangeable with any other, different serial numbers notwithstanding. But a $5 bill is not interchangeable with a $50 bill. Once again, we feel compelled to mention that this is really one of the fundamental ideas forming the basis of modern logic (through boolean logic) and modern computation (via electronic circuit design). We'll have more to say about this in the chapter on algebra.

Another important kind of relationship is the **partial order** and the **total order**. Where equivalence relations capture intuitions about equality, partial and total orders capture intuitions about inequalities such as $\leq$ ("less than or equal to"). In particular, we think of $\leq$ as being reflexive (x is always $\leq x$) and transitive (if $x \leq y$ and $y \leq z$ then $x \leq z$) but not symmetric. In fact, $\leq$ is **antisymmetric** in that if $x \leq y$ and $y \leq x$, then $x = y$.

Any such relation is called a partial ordering or partial order. Partial orders crop up in many contexts; for example, they are important in describing processes and instructions. Any baker will tell you that you need to knead the bread before you bake it and bake it before you slice it and serve it. Letting the relation $\leq$ mean "needs to be done before," you can see that $\leq$ expresses order constraints (knead $\leq$ bake, bake $\leq$ serve) on the steps in preparing food. Another example of a partial order comes from subsets. Let S be the set of all subsets of $\{a,b,c\}$ (this set of all subsets is sometimes called the **power set**, so S is the power set of $\{a,b,c\}$). The elements of S are thus

$$\{\emptyset,\{a\},\{b\},\{c\},\{a,b\},\{a,c\},\{b,c\},\{a,b,c\}\}$$

The subset relationship ($\subseteq$) defines a partial order on S. For example, $\emptyset \subseteq \{a,b\}$ and $\{a,b\} \subseteq \{a,b,c\}$. And, as the partial order properties demand, $\emptyset \subseteq \{a,b,c,\}$.

The difference between a **partial** and a **total order** is that some elements of a partial order may not be directly comparable. It doesn't matter whether you set the table before or after you slice the bread, and the Music Appreciation elective won't affect literature courses.

Similarly, $\{a,b\} \nsubseteq \{b,c\}$, but $\{b,c\} \nsubseteq \{a,b\}$, either. In such cases, we say that the elements $\{a,b\}$ and $\{b,c\}$ are **incomparable**. The two elements a and b would be **comparable** if either $a \leq b$ or $b \leq a$, but that doesn't always happen. When that does always happen—when every two elements are comparable—the relationship is called a total order. This is just one example of how complicated "numbers" are; the usual meaning of $\leq$ imposes a total order on numbers, but not everything fits naturally into the structure of a total order.

Sets and relations can be used to capture and represent lots of things that have nothing to do with "numbers," per se. In fact, this is really an area where humanists can legitimately claim to have made important contributions over the course of the last few centuries. It has probably already occured to you that some of this sounds a bit like the language used to talk about ontology (and to form and order taxonymies of various kinds). That's just one of the areas where higher mathematics bleeds over not only into a host of real-world problems and concerns, but into the loftier heights of philosophical speculation. And we're just getting started, here.

2.3 Graph Theory

Graphic designers consider Harry Beck's 1931 map of the London Underground to be one of the great masterpieces of information visualization. Before Beck, maps of railway networks tended be drawn over ordinary geographical maps, thus indicating both the distance and the angle of orientation among stations. Beck was the one who realized that in the context of a subway, such matters were

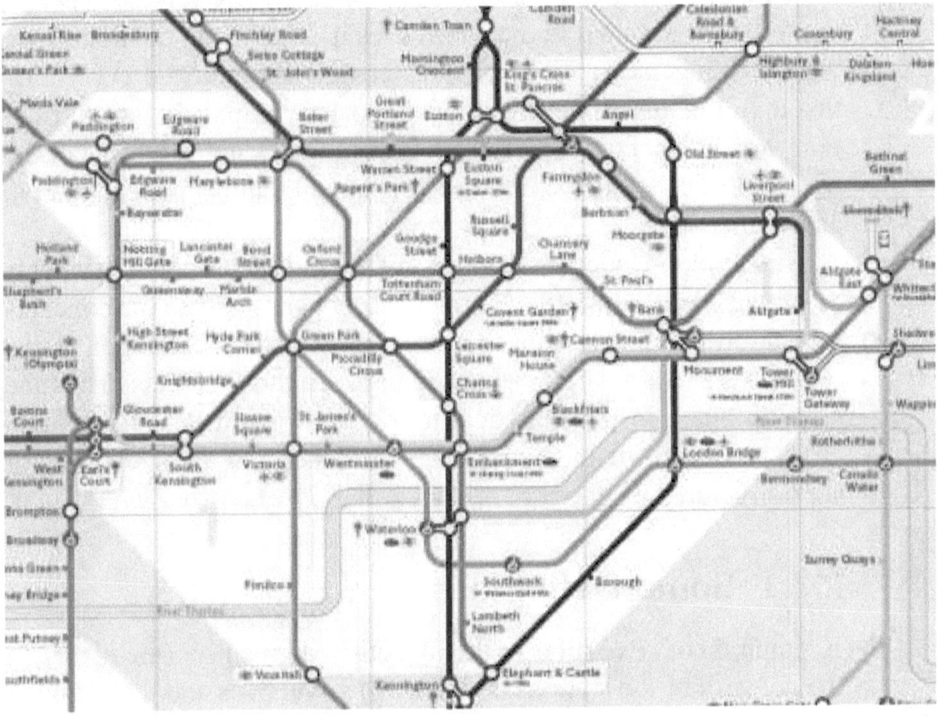

Figure 2.1: Map of of the London Underground.

mostly irrelevant. A traveler on the tube is mainly concerned with figuring out how the stations are connected in order to find out how many stops away something is and where (or whether) it is necessary to switch trains. Beck's method is now so widely imitated (nearly every subway system in the world uses it), that it can be hard to see what's so ingenious about it. Beck essentially removed everything that didn't matter. Lines lie along the major compass points and stations are mostly equidistant, even though that is not at all true of the actual physical stations. The result is a map that people can read in an instant.

The London Underground map, by boiling things down to what is essential, comes close to defining the subway as (to use the concepts we set forth in the previous section) a *set* of objects combined with a *relation* that defines "connections" between those objects. The tube map, to put it in mathematical terms, is a **graph**, and there is an entire branch of mathematics devoted to studying the properties of such structures.

2.3.1 Some Definitions

In graph theory, we refer to the objects in a given set (the nodes on the network or the stations on the map) as **vertices** and the connecting lines as **edges**. A **graph**, then, is a set V called the vertex set, together with a relation R on V defining the connections between vertices. This also implies that there is an edge set E; an edge e $(= (x,y)) \in E$ only if xRy.) Less formally, vertices are stations, while edges are trains—there is an edge from the Covent Garden vertex to the Holborn vertex along the Piccadilly line. Consider a simpler net-

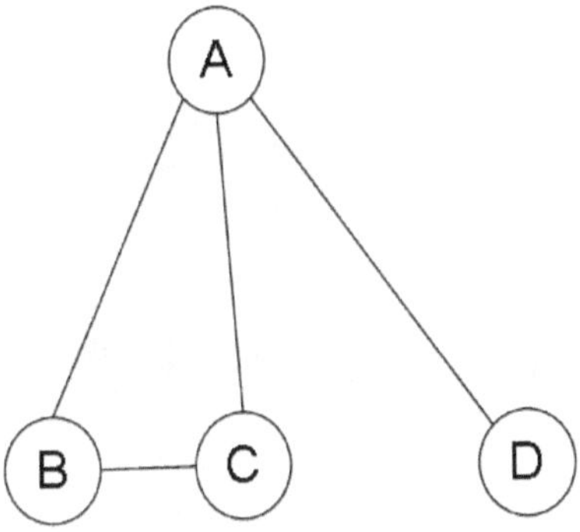

Figure 2.2: Example graph for discussion

work than that depicted on the London tube map: figure 2.2 has four vertices, so the vertex set V is $\{A,B,C,D\}$. The relation R is a subset of $V \times V$, which we can write out as ordered pairs like so:

$$\{(A,B),(B,A),(A,C),(C,A),(A,D),(D,A),(B,C),(C,B)\}$$

In this example (as in many other graphs), R is both antireflexive and symmetric. Were this a railway map, it would reflect our intuition, first, that no one would bother to build a train track that goes directly from one station back to itself with no intermediate stops, and second, that trains run in both directions.

Based on this graph, simple observation tells us, for example, that to get to D we have to go through A. But not all questions about graphs are so easily answered, and some graphs can be extremely

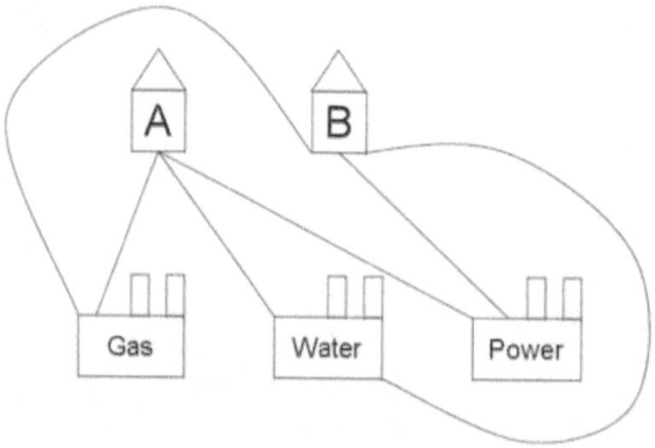

Figure 2.3: Simple planar graph

complicated—so complicated, that it makes much more sense to speak of them purely in terms of sets and relations.

2.3.2 Planar Graphs

One of the simplest questions about a graph is whether or not it can be drawn without overlapping lines. Figure 2.3 provides an example. In this figure, we have connected two houses (vertices A and B) to three utilities (Gas, Water, and Power), while allowing each utility to have its own right-of-way. We call this a **complete** graph; every house is connected directly to every utility. Since no lines cross, there is no danger that a gas technician will accidentally knock out the water to a house, and each house has its own connection, so a short in house A won't cut power to house B. The technical term for this kind of graph is **planar graph**. All of the edges fit into a plane without overlapping or crossing.

Not all graphs are planar. In figure 2.4, a third customer C has moved into the neighborhood. Notice that, as drawn, his power line crosses house B's water line. But even if you were to take a less direct path between C and the power station, it wouldn't help; in fact, even if you snake in between house A and B, you still need to cross B's gas line. If you ran C's water line across the top of the diagram instead of around the bottom, you could then run his power line across the bottom—but you'd still need to cross A's water line. As it happens, this will always be the case with three or more customers and three or more utilities, which is to say that such graphs are not planar. We won't bother to prove it, but it can be proven. As you might expect, arriving at such proofs is a matter of great interest to graph theorists.

The value of such proofs should be obvious. If you are trying to design a railway network, eliminating crossings would reduce the risk of accidents; similarly, chip designers need to minimize the number of different layers of wire to reduce chip costs. It's also a matter of enduring interest to graphic designers interested in reducing visual clutter. In each case, we want to know how few crossings we can get away with.

2.3.3 Chromatic Graphs

Lots of things don't look like graphs, but can be redrawn (or reconceptualized) as graphs. Figure 2.5 shows a (very bad) map of the western United States, along with a graph representation of the same

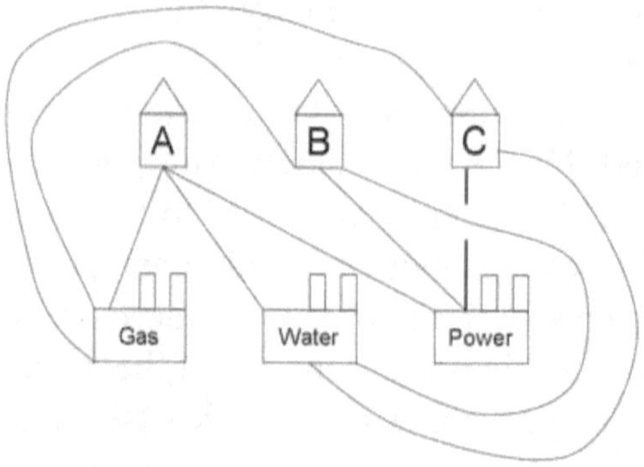

Figure 2.4: Can this graph be redrawn in a plane?

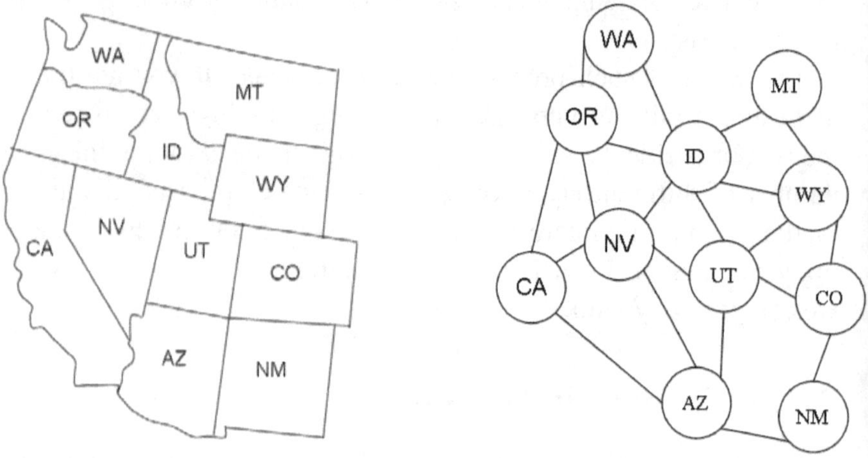

Figure 2.5: Map of Western USA and equivalent graph

picture. Now, it's considered very bad form in mapmaking to have two adjacent areas colored the same; but how many colors will it take to prevent that from happening? Obviously, we need at least three (Washington, Oregon, and Idaho need to be different colors, since they all touch), but can we do it in four? In five? In six? How many would it take to do the entire United States—or the entire world, with nearly 200 UN member-states?

We can rephrase this in terms of graphs, by simply assigning each vertex a specific color, and applying the constraint that $\forall x, y : (x, y) \in E \rightarrow color(x) \neq color(y)$. (Note the use of the function $color(x)$ to model color properties.) What is the minimum size of the range of $color(x)$?—or in plainer English, how many colors do we need for the map? This is an important problem in graph theory and in some ways still unsolved; we know the answer (although the proof is far too difficult and complicated to reproduce here) for simple (planar) graphs like figure 2.5, but not for all cases. This particular figure, like any planar graph, can be colored with four or fewer colors. In particular, we can color, for example, Washington, Montana, Nevada, and New Mexico red; Idaho, California, and Colorado green, Oregon, Arizona, and Wyoming blue, and Utah yellow. (Figure 2.6)

But a more complicated environment (perhaps a picture of the ownership of a mine or a set of subway tunnels) could take many more—and mathematicians don't always know the answer. Even in this simple case, although we know that this graph can be colored with four colors, can it be colored with only three? (We believe the answer to be "no," but you might try it yourself. Even better, see if you can prove the answer to be "no." As a hint, look closely at the

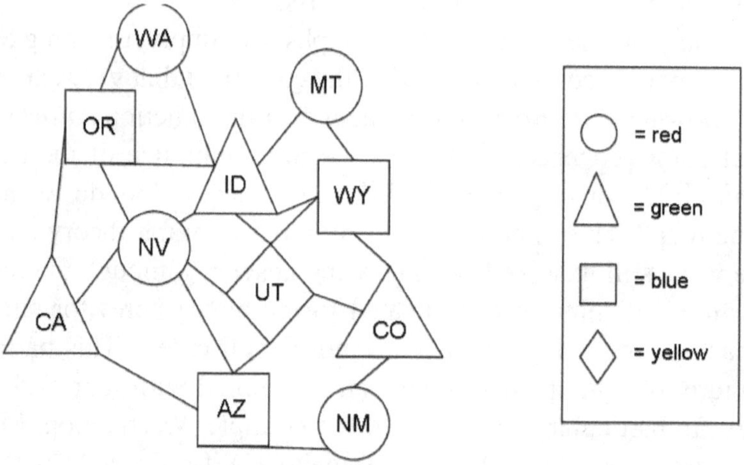

Figure 2.6: 4-colored graph of Western USA

color of Utah.)

2.3.4 Bipartite Graphs

Another special type of graph is the **bipartite** graph. A graph is bipartite whenever the vertices can be divided into exactly two groups, and every edge connects the two groups (i.e. there are no within-group edges). For example, a checkerboard is bipartite; the squares are divided into a red group and a black group, and every red square touches only black squares (and vice versa). Figures 2.3 and 2.4 are also bipartite; there is a group of utilities and a group of houses, and every edge connects a house to a utility. On the other hand, figure 2.5 is not bipartite. If you look at the triangle formed by Washington, Oregon, and Idaho, Oregon would have to be in a different group than Washington (since they're connected). But Idaho would have to be in a different group than Washington and in a different group than Oregon, so any graph division would involve at least three groups. Another way of looking at this, then, would be to notice that a bipartite graph can be colored in only two colors.

Bipartite graphs are often used to represent matching problems (sometimes called "the marriage problem"). For example, one might have a set of people P and a set of jobs J. The relation on $P \times J$ of "person p can do job j" defines a bipartite graph (as in figure 2.7); a person might be able to do any of several jobs, and a job might have several appropriate candidates, but (of course) a person can't be another person and a job can't do another job. The task of the human resources department is to find a set of people to cover all the jobs, which in turn means finding a subset of the "person p can

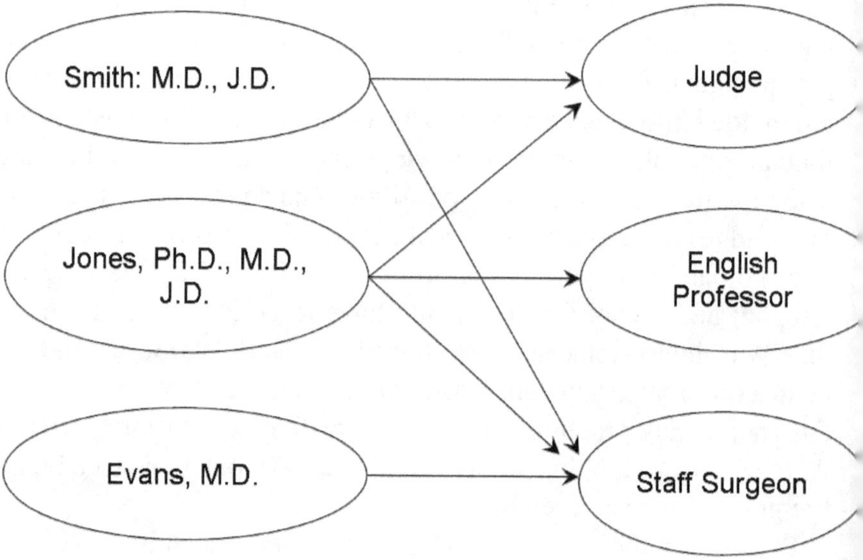

Figure 2.7: A bipartite graph describing candidates and jobs. Can you find a "marriage" to fill all the jobs?

do job *j*" relationship to represent "person *p will* do job *j*" such that all jobs are covered. Marriages, of course, can be represented in the same way; if each woman has a set of men she would be happy to marry, and each man has a set of women he would be happy to marry, a busy matchmaker or an ambitious member of the clergy has a method for pairing as many couples as possible.

A real-life example of this is the National Residency Match Program (NRMP) in the United States. With college admissions, there are so many applicants for each position that it is okay (standard practice, even) to accept more students than one expects to attend, but this doesn't work when trying to match medical school graduates to hospital residency programs. Because both medical students and slots for residents are in short supply, this kind of wastage isn't desirable—it is important to match them as well as possible. The NRMP accomplishes this using bipartite graph theory. Students and hospitals submit lists of their preferred candidates, and the resulting graph ("student *s* wants to work at hospital *h*"/"hospital *h* would like to hire student *s*") are used to define a bipartite graph, with the overall goal of maximizing everyone's happiness ("utility," as an economist might say) by placing people where they would most like to be and fit best. This program has been a well-known application of graph theory for more than fifty years, and it has been suggested that the current US college football bowl system would benefit from similar analysis. The authors (only one of whom is an actual football fan) would be happy to volunteer their services in pursuit of this goal.

2.3.5 Eulerian Graphs

A famous early problem in graph theory is the "Bridges of Königsberg," posed and solved by Leonhard Euler (1707–1783) in 1736. The city of Königsberg (now Kaliningrad) straddles the Pregolya River—including a couple of islands in the middle of the river—and in 1736 there were seven major bridges crossing it. Figure 2.8 shows a crude map of the bridges in question. A strong purist would insist that the equivalent "graph" shown below it is illegal, because there are multiple edges connecting the same points. Such a purist would be right, and you can convince yourself of this by recalling the formal definition of a graph. Such structures as that depicted in figure 2.8 are correctly referred to as **multigraphs**.[7]

According to mathematical legend, the local citizens liked to take walks about the city and wondered whether it was possible to take a walk that crossed each bridge once and only once. In graph theoretic terms, they were asking whether there exists a **path** (that is, a set of "connected vertices") that traverses every edge once and only once. Euler managed to prove that it wasn't possible, and the argument is fairly simple. Suppose, for example, that you start on the central island (B). Since there are five bridges connecting to that island, you have to go off, then on, then off, then on, then off. So if you start at B, you need to finish somewhere other than B. A similar argument shows that if you start somewhere other than B, you need to finish on B.

[7]Such notions of "purity" are obviously not meant to deny the existence of objects that don't meet the definition, but simply to ensure that the definitions are clear and unambiguous.

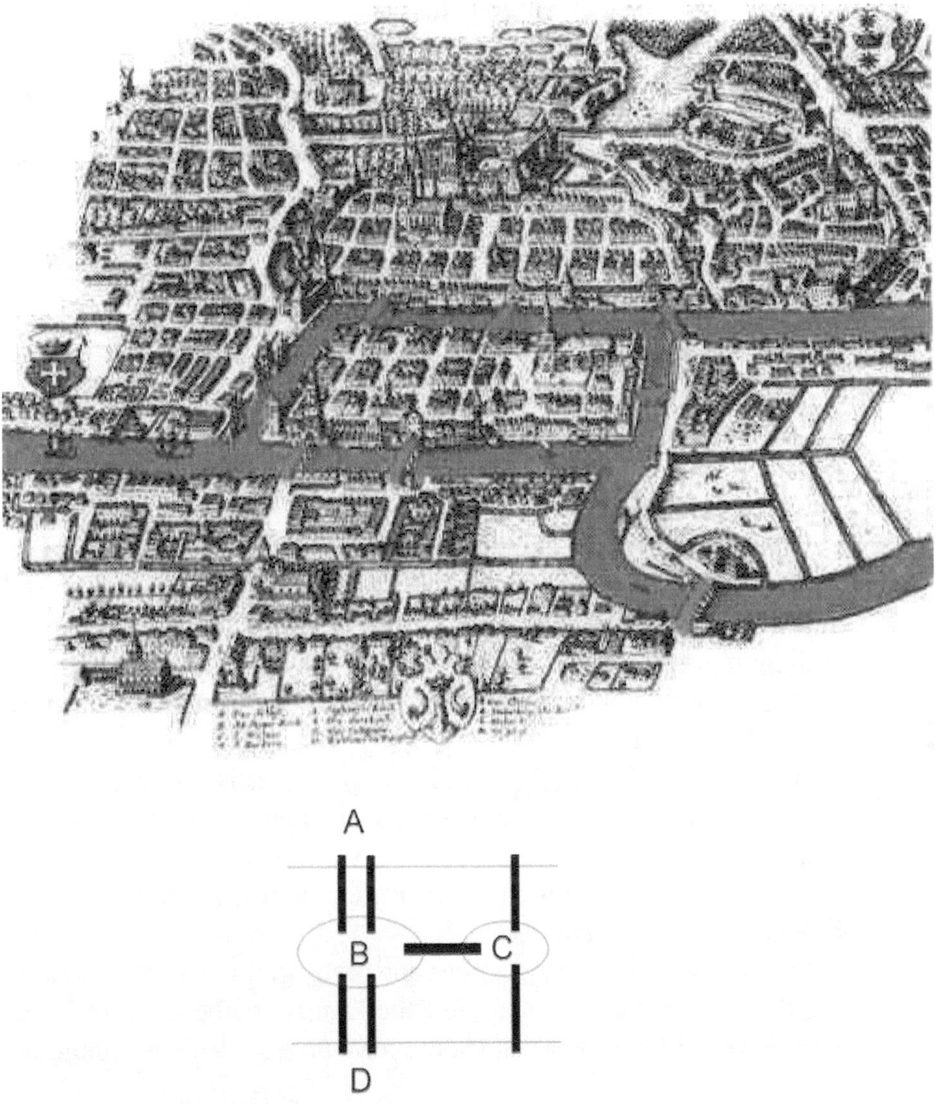

Figure 2.8: Map of Königsberg and equivalent graph

Now apply the same argument to the north shore (A). The same argument holds with three bridges as with five; if you start at A, you finish elsewhere, and vice versa. In fact, this argument holds with any odd number of bridges. It holds equally with the south bank D; if you start at D, you finish elsewhere, and if you start elsewhere, you finish at D.

But what this argument ends up saying is that you can't start (or finish) anywhere. If you start at A, you need to finish at both B (since you didn't start at B) and D (since you didn't start at D)—which is clearly impossible. If you start at B, you need to finish at both A and D. And if you start anywhere else, you need to finish at both A and B (plus possibly other places).

The key insight is that the number of vertices with an odd number of edges coming out of them matters (the technical term for this is the number of vertices of odd **degree**). If there are only two of them, the problem is solvable; you start your walk at one, and you can end at the other. If there are none, then you can walk in a circle, starting and ending at the same spot. A **cycle** is a circular path through a graph, beginning and ending at the same vertex. An **Eulerian path** includes every edge but doesn't have to begin and end at the same spot; an Eulerian **cycle** is an Eulerian path that is also a cycle, i.e. it begins and ends at the same vertex. If there is exactly one odd-degreed vertex, or more than two, the walk is impossible. A graph with an Eulerian cycle is called an **Eulerian graph**, all named after Euler and his argument.

Figure 2.9 shows a very famous and pretty graph, sometimes called the **Petersen graph** (named after the Danish mathematician Julius Petersen (1839–1910)). You should have no trouble determining for

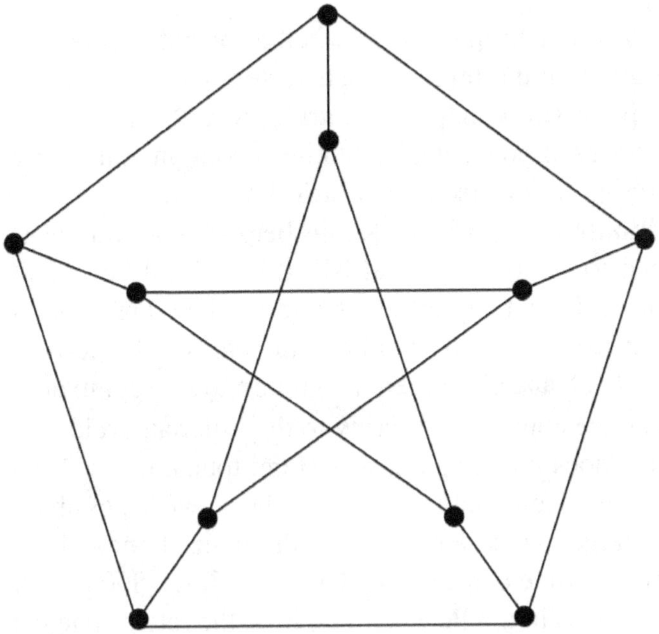

Figure 2.9: The Petersen graph—a non-Eulerian graph

yourself that the Petersen graph is non-Eulerian.

2.3.6 Hamiltonian Graphs

An obvious related question is whether or not it is possible to traverse a graph going through each vertex only once, except for the starting point (to which you return). Recall that if you're trying to traverse each edge, you're talking about an Eulerian graph. If you're trying to traverse each vertex, however, you're talking about a **Hamiltonian graph**.[8] For Königsberg, this would mean that you only need to go to the central island once, and you don't need to worry about how many different bridges there are. It's not hard to show that Königsberg is Hamiltonian; start on the north shore (A), then visit B, D, and C in order, then return to A to complete the cycle.

Despite the apparent similarity to the Eulerian cycle problem, the question, "does this graph have a Hamiltonian cycle?" is not at all easy to solve. You could, of course, check every possible ordering, but for a large graph—or even a medium-sized one—that would be impractically time consuming. There are 3,628,800 possible orderings of vertices in the Petersen graph; at the rate of one per minute, that's almost three years of full-time eight-hour days to do the work. Unlike the Eulerian cycle problem, there is no easy shortcut (or if there is, it has eluded the entire body of mathematicians).

[8]Named for the Irish mathematician William Rowan Hamilton (1805-1865), who also invented a peg-board game (called "The Icosian Game") that involves finding a Hamiltonian cycle along the edges of a dodecahedron. Will the fun never stop?

This is unfortunate, because the Hamiltonian cycle problem is important in many real-world applications. One of the most accessible, and also the most common, is the so-called **Travelling Salesman Problem**. In this problem, we assume that there is a salesman with a set of sites she needs to visit. We can imagine that the sites are vertices and that each one is connected by edges, perhaps representing the local rail or airline network. She would like to do a sales trip, starting and ending at the home office, visiting each site. At the same time, she doesn't want to waste time and money by re-visiting a site. She therefore wants to visit each vertex exactly once (or, in other words, perform a Hamiltonian cycle). For small graphs (like figure 2.2), the answer may be obvious. But how about for larger graphs? We invite you to take a few moments and try to figure out if the Petersen graph is Hamiltonian.

Right ... and that's only with ten vertices.

Just so you don't spend all night on it: No, the Petersen graph is not Hamiltonian. However, it does have two unusual properties. First, although it doesn't have a Hamiltonian cycle, it does have a Hamiltonian path. Our salesman could go through each city once and only once if she didn't need to return to the home office afterwards. Second, if you remove any vertex (and the edges running into it, of course), the resulting graph is Hamiltonian and you can do the full circle. You can verify these facts at your leisure. In the interests of full disclosure of the trivial, a non-Hamiltonian graph that becomes Hamiltonian when you delete any vertex is called **hypohamiltonian**, and the Petersen graph is the smallest hypohamiltonian graph possible.

The Travelling Salesman Problem appears over and over in domains involving logistics and optimization, and is therefore one of the most throughly studied problems in theoretical computer science. Though much is known about the problem, direct solutions (using brute force methods) tend to tax the resources of quite powerful computers even with comparatively small graphs (twenty cities, say). Some methods have been devised that produce relatively good approximate solutions, but no method exists for proving such methods to be optimal. Like so many problems in mathematics, this one is easy to state and yet extremely difficult to solve.

2.3.7 Trees

Another important category of graph is the **tree**. A **tree** is simply a connected graph that has no cycles. (A graph that has no cycles but that isn't connected is a collection of trees, and therefore referred to, appropriately, as a **forest**.) None of the graphs presented so far have been trees, but figure 2.11 shows several.

Trees, as we know, are used in many different contexts to represent relationships—especially structured or hierarchical ones. A typical geneology is usually a tree (hence the name "family tree"), although marriage between cousins can create cycles. Organization charts are usually tree-structured, as are biological taxonomies. Because of their simplicity there are a number of powerful theorems that can be proven about trees, which in turn means that a tree-structured representation is often very easy to work with.

Some examples of these theorems are listed. You may find it informative to convince yourself that they are true.

- If a tree has an edge, then it has at least two vertices of degree 1.
- There is a unique path between any two vertices of a tree.
- A tree with at least one vertex always has one more vertex than it has edges.
- Removing any edge from a tree breaks it into two disconnected components.
- Adding an edge to a tree creates a cycle.

2.4 Induction and Recursion

You'll have noticed by now that we are avoiding proofs of the various graph properties. That is in part to give you the flavor of the subject without getting too bogged down in details. But graphs provide a perfect opportunity to returning to the subject of logic and proof we explored in the first chapter.

We've already seen the mathematician's "trick" of using small concepts to build up to larger ones—sets become ordered pairs, ordered pairs become relations, and relations become graphs and trees. But the same principle can work within a concept. In particular, we can use arguments about the properties of small graphs, for example, to prove things about larger graphs, and as a result, about graphs in general. A lot of the tree properties can be demonstrated using this kind of argument.

2.4.1 Induction on Numbers

Let's shelve graphs for a second and go back to pure numbers. Consider the pattern that emerges in the following equations:

$$\begin{aligned}1 &= 1 &= 1^2\\ 1+3 &= 4 &= 2^2\\ 1+3+5 &= 9 &= 3^2\\ 1+3+5+7 &= 16 &= 4^2\\ &\vdots\end{aligned}$$

Does that pattern always hold? If you add up the first N odd numbers, do you always get N^2? The answer is "yes," as you can see from the following proof:

First, observe from the table that $1 = 1 = 1^2$. Therefore, we know that the statement is true for the smallest possible value of N. Now, let us assume that it is true that if you add up the first k odd numbers, you get k^2. That is, $1+3+\ldots+(2k-1) = k^2$. (Why $2k-1$? Well, the 1st odd number is $2 \cdot 1 - 1$. The 2nd is $2 \cdot 2 - 1$, the 3rd is $2 \cdot 3 - 1$, and so forth. So the k-th is $2 \cdot k - 1$.) Now, if that is true, then adding the next [the $(k+1)$-th] odd number gives us

$$1+3+\ldots+(2k-1)=k^2 \tag{2.18}$$

(This is our working assumption)

$$1+3+\ldots+(2k-1)+(2k+1)=$$

$$[1+3+\ldots+(2k-1)]+(2k+1)=k^2+2k+1$$

(from Equation 2.18)

$$=(k+1)^2 \tag{2.19}$$

(From high school algebra)

Therefore, adding up the first $(k+1)$ odd numbers produces $(k+1)^2$, which was to be proven. Q.E.D.[9]

In other words, if our pattern holds for any number k, it holds for the next larger one. If it holds for 100, then it holds for 101. If it holds for 50, it holds for 51. And if it holds for 1, it holds for 2. But we know that it holds for 1 (that was the first part of the proof), so it does hold for 2. And for 3, and for 4, and for 5 ... and all the way on up for any number you like.

This technique is called **proof by mathematical induction** (or simply **induction**, for short). This is similar to the logical induction

[9]The practice of signifying the end of a mathematical proof with the initials Q.E.D.—"quod erat demonstrandum," or, "which was to be demonstrated"—goes back many centuries, and is in fact a translation of a Greek phrase used by Euclid and Archimedes to signify the same thing. It's fallen a bit out of fashion among mathematicians lately, but we, being humanists at heart, like a bit of old-fashioned Latin now and again.

defined in chapter 1, but using general mathematical principles (as opposed to the properties of language). In general, an argument by induction goes something like this:

- Something is true for a small case.
- If it's true for any specific case, it's also true for a slightly larger case.

The argument thus proceeds from small cases to medium-sized ones to large ones to gargantuan ones, proving the general truth of the statement.

2.4.2 Induction on Graphs

As we said, such arguments can be created to prove many of the tree properties mentioned above. For example, we can show that *Removing any edge from a tree breaks it into two disconnected components* by showing, first, that it is true for a very small tree, and second, that if it is true for one tree, it is true for a slightly larger tree as well.

We can prove the first by simple observation. (See figure 2.12.) A tree with zero vertices almost doesn't make sense; it has no edges, and so it's not possible to remove one [case (a)]. Similarly, a tree with only one vertex has no edges [case (b)] (since that would have to be an edge from the one vertex to itself, which would be a cycle—illegal in a tree). A tree with two vertices can only have one edge [case (c)], and if you break it, the two vertices are no longer connected by anything [case (d)]. So removing an edge in a tree with two vertices breaks it into two disconnected components.

Now, assume that removing an edge from a k-vertex tree breaks it. What about a tree with $k+1$ vertices?

Let's consider a tree T with $k+1$ vertices, and designate one of the **leaves** (vertices of degree 1) as V. We will also talk about the tree T_2, which is the k-vertex tree that remains after we delete V from T. Since T_2 has only k vertices, removing any edge in T_2 will break it into two components, and since V is only connected to one other node (it's a leaf), it can't be connected to both components. Therefore, deleting an edge from T_2 would also break T into two disconnected components.

But what if we don't delete an edge from T_2? Well, the only edge that is in T, but not in T_2, is the single edge connecting V to the rest of the tree. But removing this edge would still break the tree into two components—one consisting of just V and the other of $T-V$ (which we also called T_2). So no matter what edge we remove, it breaks the graph into two disconnected components. Q.E.D. again.

In other words, we know that removing an edge from a 2-node tree breaks it. But if removing an edge from a 2-node tree breaks it, so will removing an edge from a 3-node tree. And if removing an edge from a 3-node tree breaks it, so will removing a node from a 4-node tree. And so on for as large a tree as you are prepared to contemplate.

Induction turns out to be a very powerful way to cast an argument in many different situations. You can extend the properties of small numbers to large numbers, or of small graphs/trees to large ones. But it can also be used, for example, to go from small sets to large sets, small databases to large ones, small programs to large ones, and many other things besides. In fact, it provides an easy and structured approach to questions about "the infinite." Using this type of argument, we can use two simple steps to create and validate an infinite number—literally—of conclusions.

2.4.3 Induction in Reverse

How about going the other way? Can we move from the large to the small? In many cases, we can, using a similar argument structure.

The trick is the mathematical equivalent of what heralds used to call "miss en abyme," or what is today sometimes called "the Droste effect" (named for the picture on a tin of Dutch cocoa, see figure 2.14.) By putting a picture inside a picture, you get a progression of successively smaller, but self-similar images (the box of Droste cocoa has a picture of a woman holding a box of Droste cocoa, a box with an even smaller picture of a woman holding a box of Droste cocoa, and so on). In theory, this nesting could go on forever into infinite detail, but in practical terms, the resolution of the image limits how it's actually drawn.[10]

So to draw the cocoa box, you need to draw a smaller (and less detailed, and therefore easier) version of that same cocoa box. This, fundamentally, is the method of **recursion**; solve a problem by solving a related, but smaller and easier, version of the same problem. Eventually, you come to a problem that's so small you can solve it

[10]This is also how the branch of mathematics known as **fractals** works; in particular, both the Koch snowflake and Sierpinski gasket are made by repeatedly drawing the same image only smaller and smaller. For examples, we'll defer to the many wonderful animations depicting these figures available via the Web.

just by looking at it (because the answer, mercifully, is obvious) and then build the answer to the big problem from the smaller answers. (Sometimes computer programmers will call this a "divide and conquer" approach, because it's easier to solve several small problems and combine their answers than it is to solve the big problem directly).

The idea of an Erdös number, introduced earlier, is another example of a problem that can be solved using the idea of recursion. If you want to know what your Erdös number is, you could start with Paul himself, read all of his papers, make a list of all of his co-authors, then read all of their papers and make a list of all of their co-authors, continue through their co-co-authors, and so forth.

Or, you could take a more recursive approach: Approach all of your co-authors and ask them for their Erdös numbers. The idea, of course, is that anyone who has an Erdös number of 1 knows it, since they've co-written with Paul. That's pretty obvious and easy. So if one of your coauthors says "my number is a 1," then you know that your number is a 2. But more generally, if they don't know what their number is, then they can ask their co-authors, and their co-authors can ask theirs, and so on. And whatever number they tell you—more formally, the smallest number they tell you—is one less than your number. (In our cases, Patrick's co-author Christopher Hall is a 3. This makes Patrick a 4, and by extension Steve is a 5.)[11]

[11]The people who watch over such matters (and there are such people) would probably insist that Erdös numbers only apply to people who have co-written formal, mathematical papers, and not, for example, textbooks and introductory works. Given Erdös's love for the "epsilons" (the conventional symbol for a small, positive quantity and the term he routinely used to designate children),

We therefore have a recursive definition of Erdös numbers. If you have written a paper with Paul, your number is a one. Otherwise, your number is one more than the smallest number of any of your co-authors.

As another example, consider a simple safe with a programmable numeric keypad. One way of getting into such a safe is simply to try all possible combinations until you get in. If the combination is only one digit, there are ten possible combinations, and we can probably get in in a few seconds. If the combination is two digits long, there are a hundred possibilities, and it will take longer. But what if there are four digits? Or five? Or ten?

To solve this, notice that we can break up the combination like a phone number (think of the difference between 345-8082 and 3458082); if we have x + 1 digits in the combination, then that's the same as if we had a 1-digit combination followed by an (x)-digit combination (3-458082). There are obviously ten possibilities for a one digit combination, 0,1,2,3,4,5,6,7,8,9. Perhaps equally obviously, if there are N combinations for an x-digit combination, then we can combine any of the possible first digits with any of the possible ending combinations. This is simply the Cartesian product defined earlier, and it produces a set of 10N possible (x + 1)-digit combinations.

---

we feel confident that anyone who devotes their life to teaching mathematics to the young is a friend of Paul.

We thus have the following set of relationships:

$$\begin{aligned} N(1) &= 10 \text{ (There are 10 possible 1 digit combinations)} \\ N(x+1) &= 10 \cdot N(x) \\ &\quad \text{(Adding a digit produces 10 times} \\ &\qquad \text{as many combinations)} \end{aligned}$$

So what is $N(5)$ (the number of 5-digit combinations)? Well, by the relationship above:

$$\begin{aligned} N(5) &= 10 \cdot N(4) \\ N(5) &= 10 \cdot 10 \cdot N(3) \\ N(5) &= 10 \cdot 10 \cdot 10 \cdot N(2) \\ N(5) &= 10 \cdot 10 \cdot 10 \cdot 10 \cdot N(1) \\ N(5) &= 10 \cdot 10 \cdot 10 \cdot 10 \cdot 10 \\ N(5) &= 10 \cdot 10 \cdot 10 \cdot 100 \\ N(5) &= 10 \cdot 10 \cdot 1000 \\ N(5) &= 10 \cdot 10000 \\ N(5) &= 100000 \end{aligned}$$

The important thing to notice here is that every statement involves a *restatement* of the problem we're trying to solve that includes some smaller version of the problem itself. $N(4)$ is part of the definition of $N(5)$, while the definition of $N(4)$ involves the definition of $N(3)$. But as we go "down" through the problem, we find that the N eventually becomes the simplest problem of all ($N(1) = 10$). This then allows us to go back "up" to the hard problem (which could be very hard indeed).

At one combination per second, this would take just under 28 hours to check all the combinations. So if someone has uninterrupted access to your safe for a weekend (and a lot of coffee), they could get in, but they probably can't do it in a lunch hour. (If you need more security, a six-digit combination would take a couple of uninterrupted weeks, and a seven-digit combination would take several months.) We could just as easily figure out what $N(1000)$ is, but that would take a lot of paper to write out, and a meaninglessly large time (trillions of years) to pick. The definition, however, is easy to understand; the larger number is just ten times the next smaller. $N(1000)$ is simply $10 \cdot N(999)$.

An important function often defined in this way is the **factorial** function; as we will see in the following section, it has a lot of applications in combinatorics and later in probability and statistics.

We can express the factorial function (written using the symbol !—while resisting the urge to say the number louder) as,

$$\begin{aligned} 1! &= 1 \\ 2! &= 2 \cdot 1 \\ 3! &= 3 \cdot 2 \cdot 1 \\ 4! &= 4 \cdot 3 \cdot 2 \cdot 1 \\ &\vdots \end{aligned}$$

But what's the actual definition? Well, it's

$$\begin{aligned} 1! &= 1 \\ n! &= n \cdot (n-1)! \end{aligned} \tag{2.20}$$

So 5! is $5 \cdot 4!$, which can be broken down further into $5 \cdot 4 \cdot 3!$ until we get down to 1! which is 1.

Factorials are very useful, particularly when it comes to calculating the possible number of ways in which something can happen. For example, let's say that four teams are competing in a tournament. Excluding ties, how many different ways can they finish? Well, any of the four teams can win. And any team except the winner can place second, so once a winner has been named, there are three possible runners-up. Anyone except the winner and runner-up can place third, so there are two possibilities, and of course, the one remaining team must have placed last. So there are $4 \cdot 3 \cdot 2 \cdot 1$, or 4!, possible outcomes (see table 2.1). This same argument would apply to any number of teams; if there are 100 teams, then there are 100 possible first-place finishers, 99 possible second-place, and 100! possible total outcomes.

So how many possible outcomes are there in a race where no one competes? Well, the silly result of that silly situation is always the same, right? Therefore, 0! is defined as 1—with important consequences, as you'll see in the next section.

First	Second	Third	Fourth
1	2	3	4
		4	3
	3	2	4
		4	2
	4	2	3
		3	2
2	1	3	4
		4	3
	3	1	4
		4	1
	4	1	3
		3	1
3	1	2	4
		4	2
	2	1	4
		4	1
	4	1	2
		2	1
4	1	2	3
		3	2
	2	1	3
		3	1
	3	1	2
		2	1
Four choices	Three choices	Two choices	One choice

Table 2.1: 24 possibilities for a four-runner race

2.5 Combinatorics

We have drifted, almost imperceptibly, into the discipline of **combinatorics**. Combinatorics is the study of counting things, like the number of possible finishes in a race or the number of "combinations" on a safe lock. Although this may seem a rather dull area, it (like logic and set theory) forms a crucial foundation for many other math subfields. Fortunately, it's both crucial and accessible—the foundations and principles are easy to understand.

The most basic principle of combinatorics is the idea of the Cartesian product; the idea that if you have a set F of possible first events and a set S of possible second events, then the set of possible event pairs is simply $F \times S$. And if there are f different ways the first thing can happen (in other words, $|F| = f$) and s ways for the second ($|S| = s$), then there are $f \cdot s$ different ways in total. For example, license plates in Pennsylvania used to have three (capital) letters followed by three digits. There are 26 letters and 10 digits, and hence $26 \cdot 26 \cdot 26 \cdot 10 \cdot 10 \cdot 10 = 17,576,000$ possible license plates. Since the population is about 12 million, this meant that there were eventually not going to be enough plates to go around. At five to ten million car registrations per year, the supply would run out in about two years. The state, therefore, added a fourth digit to the license numbers, so there are now $26 \cdot 26 \cdot 26 \cdot 10 \cdot 10 \cdot 10 \cdot 10 = 175,760,000$ different plates (enough for nearly twenty years).

A slightly harder case emerges when the events are not completely independent. If the first letter on your license plate is a 'P,' nothing prevents the second from being a 'P' as well. But the person who came in first in a race cannot come in second as well. This case yields the factorial we saw in the previous section. If there are n possibilities for the first element, there are only $n-1$ for the second, and so forth.

This kind of problem is called a **permutation**: a rearrangement of a collection of objects. We could ask, for example, how many different ways there are to rearrange a set of seven crayons, or how many different seating arrangements there are for a dinner party of twelve people. In general, if there are n items that need to be placed, there are $n!$ different permutations. But things can get tricky if not all objects need to be placed.

Returning to the race example: Let's suppose that we have a race with 20 competitors. Discounting ties, there are 20! different possibilities for the total order of finishing. But in this case, we are giving medals only to the top three finishers, and we just want to know how many different ways there are to distribute the medal.

We can use a similar argument to the one from the previous section: any of 20 people could have won the gold medal. Anyone except the gold medalist—any of 19 people—could have taken the silver, and anyone except the gold or silver medalist—any of 18 people—could have taken the bronze. So there are $20 \cdot 19 \cdot 18$ possible medal sets; only 6,840 instead of the trillions of complete lists.

In general, mathematicians will speak of the "number of permutations of r out of n objects," and since that's a mouthful, they will use the notation $P(n,r)$. So our example shows that $P(20,3)$ (the number of 3-permutations out of 20) is $20 \cdot 19 \cdot 18$. In general, the formula used is

$$P(n,r) = \frac{n!}{(n-r)!} \tag{2.21}$$

Mathematically, you can see how this works; we're simply dividing out the number of irrelevant combinations. If Smith, Jones, and Evans are the three medalists, then there are 17 non-medalists who can arrange themselves in 17! different ways (that we explicitly ignore, because we don't care about them). But this is true no matter who the three medalists are: there are still 17! arrangements of the nonmedalists that we ignore. We therefore find that

$$\begin{aligned} P(20,3) &= \frac{20!}{(20-3)!} \text{(From Equation!2.21)} \\ &= \frac{20!}{(17)!} \\ &= \frac{20 \cdot 19 \cdot 18 \cdot 17 \cdot \ldots \cdot 1}{17 \cdot \ldots \cdot 1} \\ &= \frac{20 \cdot 19 \cdot 18 \cdot 17!}{17!} \\ &= 20 \cdot 19 \cdot 18 \end{aligned}$$

as required.

Notice that if $r = n$, this also produces the result we've used before. Since 0!=1, $P(n,n) = \frac{n!}{(n-n)!} = \frac{n!}{0!} = \frac{n!}{1} = n!$.

The most difficult case of all is the one where ordering doesn't matter. If the only purpose of the race is to determine the three fastest people (who will then advance to the finals), all that matters is who those three are, not what order they finished. Because order no longer matters, we can permute the winning triple in any of 3! different ways while still getting the same answer. In turn, this means that 3! different permutations only count as one "real" answer, so the total number of permutations needs to be divided by 3! to get the true number.

Mathematicians speak of this as the "number of **combinations** of r out of n objects." There are two fairly common notations for this idea:

$$C(n,r) = \binom{n}{r} = \frac{n!}{(n-r)!r!} \tag{2.22}$$

So the number of different finishing permutations is $20 \cdot 19 \cdot 18$ or 360. But the number of different finishing combinations is $\frac{20 \cdot 19 \cdot 18}{3 \cdot 2 \cdot 1}$, or only 120.

As we said in the beginning, discrete mathematics is a great source of games and puzzles. So as final example of how this works, let's play a little poker.

The best possible hand in poker is the royal flush, which consists of an A, K, Q, J, and 10 in the same suit. There are exactly

$$\binom{52}{5} = \frac{52!}{47!5!} = \frac{52 \cdot 51 \cdot 50 \cdot 49 \cdot 48}{5 \cdot 4 \cdot 3 \cdot 2 \cdot 1} = 2,598,960 \tag{2.23}$$

ways to be dealt a poker hand of five cards. Of those ways, exactly one is a royal flush in spades, so the odds of your getting such a hand are about two-and-a-half million to one against.[12] Exactly four of the two million plus hands are royal flushes at all (one per suit), so the odds of getting any royal flush are 1 in $\frac{2{,}598{,}960}{4}$ or 1 in 649,740. On the other hand, if all you want is a straight flush (cards in suit and sequence, but not necessarily starting at an ace-high), there are 9 ways of getting that in each suit, or thirty-six ways in total. The odds against that are a mere 1 in 72,193.33.

A flush is simply all five cards of the same suit. There are thirteen hearts in a standard deck from which we need to choose any unordered set of five, so there are

$$\binom{13}{5} = \frac{13!}{5!8!} = 1287 \tag{2.24}$$

hands that are all clubs, or $4 \cdot 1287 = 5148$ hands that are all the same suit. The chances of being dealt a flush in clubs are therefore 1 in $\frac{2{,}598{,}960}{1287}$, and the chance of being dealt a flush in any suit are 1 in $\frac{2{,}598{,}960}{5148}$, or about 1 in a bit over 500.

A more complicated example would involve a full house, defined as a combination of three of a kind and a pair of another kind. What are the odds against this? How many possible ways are there to be dealt a full house?

Let's start by looking at a specific example—say, for example, three queens and two eights. There are four queens (and eights) in total, so there are $C(4,3) = 4$ different ways to be dealt three

[12] We encourage you to use these techniques to calculate the odds of winning your standard "Mega Millions"-style lottery. We feel confident that you will never play again.

queens and $C(4,2) = 6$ different ways to get a pair of eights. By the Cartesian product principle, there are thus $4 \cdot 6$ or 24 different hands with three queens and two eights. But there are 13 different ranks that we could have been dealt instead of queens, and 12 different ranks for the pair. There are therefore $24 \cdot 13 \cdot 12$ or 3744 different full houses.

Therefore, the odds of being dealt queens over eights are 1 in $\frac{2{,}598{,}960}{24}$, while the odds of being dealt a full house at all is 1 in $\frac{2{,}598{,}960}{3744}$, just about 700:1 against.

This simple calculation shows why the hands are ranked as they are. A royal flush is the rarest and for that reason the most valuable. A full house is not nearly as rare as a royal flush, but is less common than a simple flush, and therefore beats it.

Once again, we are crossing into another subject: probability and statistics, the subject of a later chapter. Like discrete mathematics, probability is full of real-world examples and abundant opportunities for practical application. But we would instead like to delve into a subject that in subtle ways undergirds all of mathematics. We are speaking, of course, of algebra.

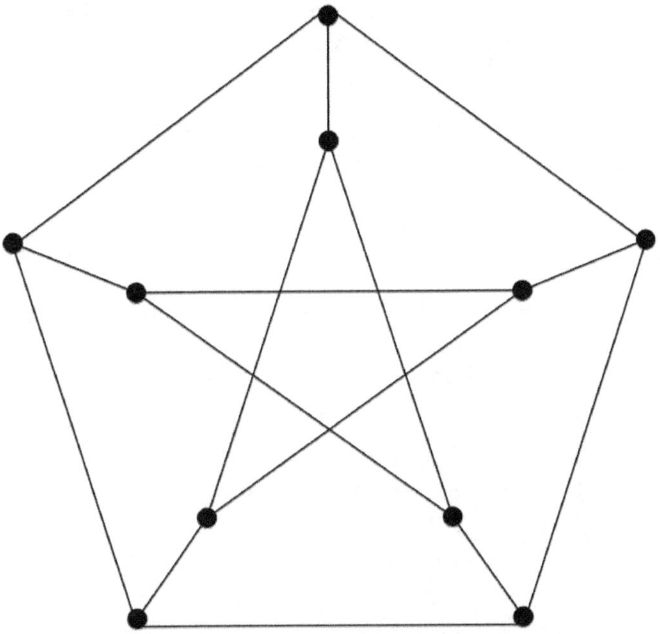

Figure 2.10: The Petersen graph—is it also non-Hamiltonian?

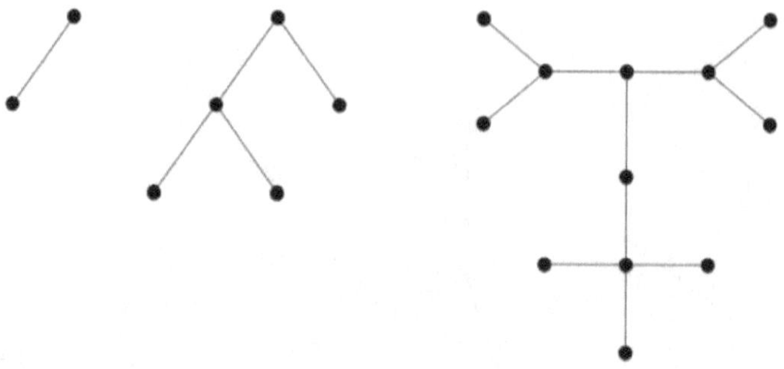

Figure 2.11: A forest. Each component is a tree

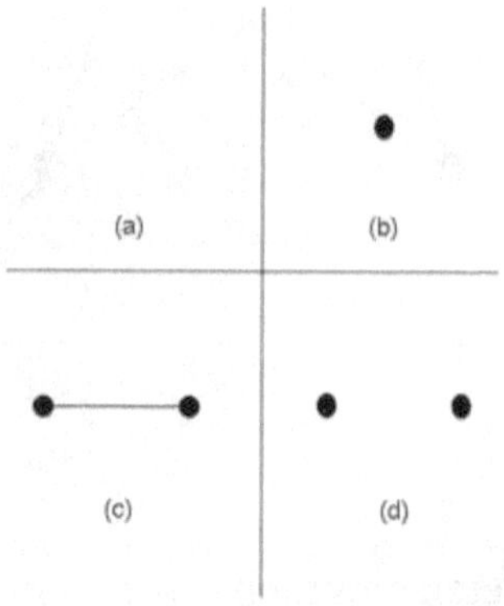

Figure 2.12: Removing edges from small trees

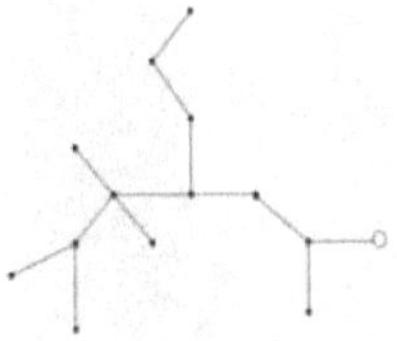

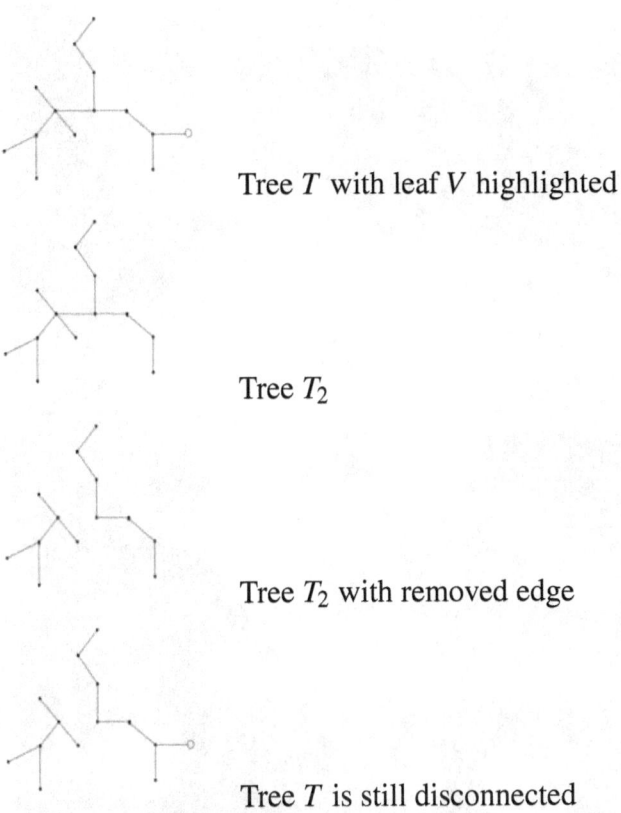

Tree T with leaf V highlighted

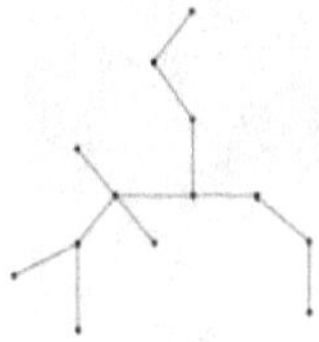

Tree T_2

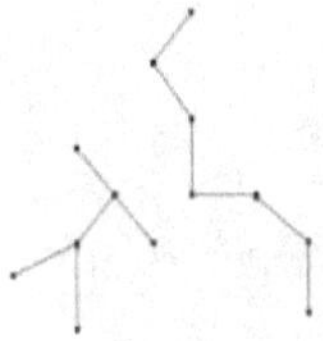

Tree T_2 with removed edge

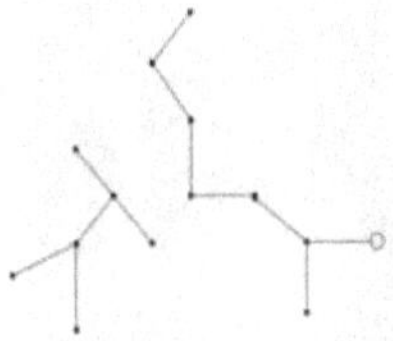

Tree T is still disconnected

Figure 2.13: Removing edges from larger trees

Figure 2.14: The Droste Effect.

3 Algebra

"When I considered what people generally want in calculating, I found that it is always a number."

So begins *The Compendious Book on Calculation by Completion and Balancing* by the Persian mathematician Muḥammad ibn Mūsā al-Khwārizmī (c. 780–c. 850)—the book from which we get the word "algebra" (in Arabic, *al-ğabr*: "restoration"). In the introduction to his work, al-Khwārizmī proposes to confine his discussion "to what is easiest and most useful in arithmetic, such as men constantly require in cases of inheritance, legacies, partition, law-suits, and trade, and in all their dealing with one another" [3]—a sentiment that continues to echo in the words of high-school teachers eager to make an occasionally rather dull subject seem relevant to the young.

The methods themselves were hardly new. The roots of algebra, as an algorithmic (the word "algorithm," by the way, derives from the name of al-Khwārizmī himself) method of calculation, are to be found not only in Greek mathematics, but in hundreds of Babylonian tablets going back to nearly four-thousand years ago. Al-Khwārizmī's contribution, though, extends far beyond the mere matter of how to calculate. The historians R. Rashed and Angela Armstrong write:

> Al-Khwarizmi's text can be seen to be distinct not only from the Babylonian tablets, but also from Diophantus' *Arithmetica.* It no longer concerns a series of problems to be resolved, but an exposition which starts with primitive terms in which the combinations must give all possible prototypes for equations, which henceforward explicitly constitute the true object of study. On the other hand, the idea of an equation for its own sake appears from the beginning and, one could say, in a generic manner, insofar as it does not simply emerge in the course of solving a problem, but is specifically called on to define an infinite class of problems.

It would be difficult to overstate the importance of this shift in the understanding of algebra. With al-Khwārizmī, algebra goes from being a system of rote methods for solving numerical problems to an investigation of the structures and relationships that govern mathematics itself. That, in fact, is what contemporary mathematicians mean when they speak of algebra (sometimes referring to it as Abstract Algebra or Modern Algebra to distinguish it from its grade-school cousin). With al-Khwārizmī, algebra takes its rightful place as the mathematics of mathematics.

A forbidding subject, to be sure. But the ideas that govern "algebraic thinking" in this latter, more philosophical sense reappear in every branch of mathematics without exception.

We are stepping up the complexity of our discussion quite a bit in this chapter; you are unlikely to have encountered these ideas outside of a formal class in college-level mathematics, and so much of it may be entirely new. We recommend that you proceed slowly, and resist the urge to become discouraged if things don't seem immediately obvious. In return, we will offer the bargain that many a high-school teacher has presented to his or her charges: this will pay off later.

3.1 Structure and Relationship

It is worth our while to linger over the idea of "structure and relationship" for a moment. Consider (at the risk of creating an even greater sense of trepidation) the "analogy" problems that often appear on standardized tests:

United States : Congress :: United Kingdom : ?

The answer is assuredly "Parliament," and we could perform the same logical deduction with respect to the Israeli Knesset, the Russian Duma, the Irish Houses of the Oireachtas, and with the governing bodies of dozens of nation states. But what, precisely, is the relationship we are talking about? Is it possible to generalize this kind of relationship beyond the narrow realm of comparative political structures—perhaps into areas that have nothing at all to do with politics or nations? What are the properties that persist in other structures that are superficially similar, but not quite the same?

Another example of this can be seen in the "sudoku" puzzles that have recently become popular in much of the world. The basic structure is a nine-by-nine grid, in which solvers are asked to place the digits 1 through 9 such that no row, column, or three-by-three subsquare has more than one of each digit. We provide you with one

4	8		5		7		9	2
	1		2		3		6	
		7				9		
1				7				4
		8				2		
	7		1		9		5	
5	9		7		2		8	3

Figure 3.1: Sample sudoku puzzle in standard format

example here; a well-stocked bookstore should provide you with thousands if needed.[1]

When you inspect a sudoku puzzle closely, it becomes apparent that even though there are numbers in the grid, it is not actually a "number" puzzle at all. It is really a puzzle that involves coming up with patterns that fulfill certain specified conditions. Instead of using numbers (as in figure 3.1), we could pose the same problem with anything for which we could produce nine different variations.

[1]The solution is given at the end of the chapter.

3B	CF		P		RF		LF	2B
	1B		2B		SS		C	
		RF				LF		
1B				RF				3B
		CF				2B		
	RF		1B		LF		P	
P	LF		RF		2B		CF	SS

Figure 3.2: Sudoku puzzle for baseball fans

We could use the letters A–I, the positions on a baseball team (figure 3.2), or the nine Muses (figure 3.3).

This suggests that there is at least one structural property that is entirely independent of the specifics of sudoku as commonly played. Are there other such properties? Do those properties change as the grid gets larger? Would the essential properties be different if played amid a set of interlocking triangles or on the surface of a sphere? Are there other games that are reducible to the essential structure of sudoku (despite seeming entirely unrelated)?

Insight into essential relationship—the quest for those things that

Eu	Th		Me		Te		Ur	Cl
	Ca		Cl		Er		Po	
		Te				Ur		
Ca				Te				Eu
		Th				Cl		
	Te		Ca		Ur		Me	
Me	Ur		Te		Cl		Th	Er

Figure 3.3: Sudoku puzzle for Greek myth fans (the nine muses are named Calliope, Clio, Erato, Euterpe, Melpomene, Polyhymnia, Terpsichore, Thalia, and Urania).

do not vary among outward forms—has often been the source of great breakthroughs in mathematics. René Descartes (1596–1650) provided one of the more striking examples in his work with the coordinate system that bears his name (Cartesian coordinates). Before Descartes, geometry and arithmetic were studied separately, and the properties of geometric objects like ellipses and parabolas were defined as geometric properties of points and distances. Descartes's great insight was to think of these objects as *sets of numbers,* which in turn allowed him to describe their properties via arithmetic. This made it possible, for example, to determine where two curves intersected by describing the equations of the two curves and solving for the numbers incorporated into both. Instead of drawing lines (and likely making mistakes along the way), we became able to solve geometric problems using arithmetic methods with clear, demonstrable solutions.

Realizing that one could describe the properties of space in terms of the properties of numbers ultimately entailed an intuition about the structure that underlies both geometry and arithmetic, and which has some of the same form as the questions we just posed about analogies and sudoku. In the case of Descartes's insight, the effect was a revolution in mathematical understanding. We refer to the subfield created by these insights, accurately and appropriately, as "algebraic geometry."

3.2 Groups, Rings, and Fields

3.2.1 Operations

Conceiving things algebraically usually involves an attempt to discern the fundamental properties of a given set of mathematical objects. One of the most essential instances of such generalization in algebra is the idea of an **operation**, which is simply a procedure that produces an "output" value when given one or more "input" values.

This is closely related to the idea of a function, as introduced in the previous chapter. Suppose we have a set S, and define $S \times S$ as the set of ordered pairs of elements in S. We can create a function from this set $S \times S$ to the set S itself, thus defining a "binary operation" on this set.

For example: Addition takes two numbers and returns a third. Given the pair of numbers (4, 3), the operation +(4,3) [normally written as 4+3] is of course 7. But that's not the only example of a binary operation. Subtraction is also a binary operation, as is multiplication, and so forth. But we could also define the *max* operation as a binary operation that returns the larger of the two numbers, so $max(4,3)$ would be 4. It is also possible to define binary operations on things other than numbers. We could let S be "the set of muses," in which case a sample drawn from $S \times S$ would be one of the pairs (say, Clio and Terpsichore). We could then define a function f as "the first of the pair in alphabetical order." In this case f(Clio, Terpsichore) would be Clio. Similarly, if we define S as the set of dates, and f as the "the latter of the pair", then $f(5/12/2001, 8/6/2014)$ would be in summer 2014. (Oddly enough, it would be in summer 2014 regardless of whether you read them as European-style or American-style dates.)

We could generalize the *max* function to take more than two arguments, so $max(3,5,7,9)$ would be 9. But since it now takes more than two arguments, it ceases to be a binary operation. Alternatively, we could define an operation that goes from one set to a different set—for example, an operation that takes a pair of dates and returns the number of days between them. But since "a number of days" isn't itself a date, that would not fit the definition we gave above.

For reasons that will become apparent, we usually use a slightly different notation for these functions, putting a funny symbol (we'll use $\circ$) between the two elements of the pair instead of surrounding the pair with $f()$, like so: Clio $\circ$ Terpsichore. So with the definitions above (where we take the alphabetically earlier of the pair), Clio $\circ$ Terpsichore = Clio.

From the discussion above, it should be obvious that if the two arguments to the function are in S, the function value itself will also be in S. In this case, we say that that this function $\circ$ is "closed" over S (i.e. exhibits the **closure** property). This makes intuitive sense and is an important property of many kinds of numbers. For example, if you add two positive numbers, you get a positive number—the operation "addition" is closed over the positive numbers. If you multiply two integers, the result is an integer—the operation "multiplication" is closed over the integers. However, division is not—you can divide one integer by another and get a result that isn't an integer (like $\frac{2}{3}$).

From the closure properties, it follows that the function can be **composed**—that is, applied several times in succession. For example, and still using the notation we've been developing, the expression

$$(Clio \circ Terpsichore) \circ Euterpe$$

is (because Clio ∘ Terpsichore = Clio) the same as

$$Clio \circ Euterpe$$

and has the same final result : Clio.

In general, any closed function can be composed like this. But in addition, the ∘ function we've been describing has another important property: it doesn't matter how we group the elements before composing the function—a property known generally as **associativity**. (A function with the associativity property is said to be **associative**.)

For example (using parentheses to make matters a little clearer), we can consider the following three ways of grouping an expression.

(((Clio ∘ Euterpe) ∘ Terpsichore) ∘ Erato)
((Clio ∘ Terpsichore) ∘ Erato)
(Clio ∘ Erato)
Clio

((Clio ∘ (Euterpe ∘ Terpsichore)) ∘ Erato)
((Clio ∘ Euterpe) ∘ Erato)
(Clio ∘ Erato)
Clio

(Clio ∘ (Euterpe ∘ (Terpsichore ∘ Erato)))
(Clio ∘ (Euterpe ∘ Erato))
(Clio ∘ Erato)
Clio

Whether from the left, the right, or the inside-out, the answer will always be the same (for this particular definition of $\circ$). If you reach back very far into your memory of school algebra, you may recall that addition is also associative: $3+(4+5)$ is the same as $(3+4)+5$. However, subtraction is not: $3-(2-1)=2$, while $(3-2)-1=0$.

3.2.2 Groups

The set and binary operation $\circ$ fulfill most, but not all, of the criteria for being a **group**, which (speaking formally) is "a model of a closed operational structure." Stripped of the technical language, a group is a set of items and a structured way of getting from one item to another. This could be something like adding two numbers and getting a third or rearranging diners at a table and getting a new seating arrangement.

There are two other criteria.

The first is the existence of a so-called **identity** element. This is an element that we name e such that for any element a, $e \circ a$ or $a \circ e$ is simply a itself. It's an element that doesn't change anything, like adding zero or multiplying by one. (In fact, the element is often written as 0 or 1, as we'll see later).

In the $\circ$ operation as we've defined it, Urania is the identity element. For example, Polyhymnia $\circ$ Urania is the same as Urania $\circ$ Polyhymnia, namely Polyhymnia herself. (If you want to see why, notice that Urania is, alphabetically, the last of the Muses.)

The second is the existence of an inverse element. For any element a, the inverse of a (sometimes written a', $\bar{a}$, $-a$, or a^{-1}) is the element such that $a \circ a^{-1} = e$, the previously-defined identity.

Our system involving the muses is not a group, precisely because not all elements have inverses. (In particular, notice that Calliope $\circ$ *ANYTHING* = Calliope, because Calliope is the first muse alphabetically. Hence there is no element $Calliope^{-1}$ such that Calliope $\circ$ $Calliope^{-1}$ = Urania.)

A group, then, is simply a set S and a function $\circ$ such that

- The function is closed over S, which is to say that an element of S $\circ$ an element of S is another element of S.
- The function $\circ$ is associative, which is to say that $(x \circ y) \circ z$ is always equal to $x \circ (y \circ z)$.
- S has an identity element e such that $e \circ x = x \circ e = x$ for all x in S.
- Every element x in S has an inverse x^{-1} such that $x \circ x^{-1} = e$ (the identity element).

Now, why do we care? Why are we doing this? In essence, by defining a thing called a group, we are generalizing (and capturing most of the key properties of) addition. Instead of talking about what happens with particular numbers, we are providing a robust description of the structure and set of relationships that characterize the "system" that the numbers represent. What's more, any system that can be represented by a group is one in which our intuitive understandings of (the operation known as) addition can be usefully deployed.

If the muses as defined do not form a group, what does? Classical examples of groups include things like the set of integers (with respect to addition), the set of real numbers (with respect to addition), and so forth. An interesting non-classical example is simply a clock—or more accurately, time as represented on a clock.

Clock arithmetic does not follow the rules of ordinary arithmetic. Consider the following story problem: *Susan has a curfew of 10 o'clock. If she is fifteen hours late, what time does she get home?* She doesn't get home at 10+15=25 o'clock; no such time exists.[2] However, we can use the rules of the group as applied to clock arithmetic to figure out that she gets home at 1:00 the next day.

Notice, first, that the set of times is well-defined, and second, that adding any time to any other time still yields a time. Second, notice that adding 12 hours to a time does not change the time (as expressed on the clock). More formally, S in this case is a time or duration, and $\circ$ is simply adding two times/durations and subtracting 12 as necessary when one passes 12 o'clock. This is clearly closed and composable. Furthermore,

- $\circ$ is associative, since adding durations can be done in any order
- 12:00 is an identity element that does not change the time; $x \circ 12:00 = x$ as needed.

[2]One of the more poignant recent examples of this property occurred during the Y2K scare at the turn of the millennium. The problem, essentially, was that some clocks on computer systems were set up as hundred-year clocks, with the result that 1999+1=1900.

- Any number of hours after 12:00 has an inverse in that many hours before. We can see, for example, that the inverse of 3:00 is 9:00 and vice-versa, since 3:00 $\circ$ 9:00 = 12:00.

If we understand addition of integers, then by extension we understand the extension of addition to clock arithmetic and can apply our understanding in a new domain. In fact, the idea of "clock arithmetic" crops up in many areas of science, including physics, computer science, and cryptography.

Another key insight relates to the idea of the group structure itself. Once we have identified a particular group structure, it may be possible to relate that structure to the structure of other groups. For example, one group may be **isomorphic** to another if there is a mapping between the two groups that preserves the group structure.

Another example: Any ten year old is familiar with the idea of a "remainder;" when you divide one number by another number, they may not go "evenly" (for example, 22 divided by 4 is 5, with "remainder" 2). Mathematicians have formalized this notion as the **modulus** operator, written "**mod**." 22 mod 4, for example, is 2—because the remainder is 2 when you divide 22 by 4. This simple operation turns out to be useful in a variety of contexts, precisely because it can model math in restricted domains such as on the face of a clock.

For example, letting S' be the set of numbers {0,1,2,3,4,5,6,7,8,9, 10,11}, table 3.1 shows how the numbers can be added (mod 12). 5 + 8, for example, becomes not 13, but 1, because 13 mod 12 is 1. This table can be seen to describe the same relationship as clock arithmetic, with one single change: the number 12:00 is described

+	0	1	2	3	4	5	6	7	8	9	10	11
0	0	1	2	3	4	5	6	7	8	9	10	11
1	1	2	3	4	5	6	7	8	9	10	11	0
2	2	3	4	5	6	7	8	9	10	11	0	1
3	3	4	5	6	7	8	9	10	11	0	1	2
4	4	5	6	7	8	9	10	11	0	1	2	3
5	5	6	7	8	9	10	11	0	1	2	3	4
6	6	7	8	9	10	11	0	1	2	3	4	5
7	7	8	9	10	11	0	1	2	3	4	5	6
8	8	9	10	11	0	1	2	3	4	5	6	7
9	9	10	10	0	1	2	3	4	5	6	7	8
10	10	11	0	1	2	3	4	5	6	7	8	9
11	10	0	1	2	3	4	5	6	7	8	9	10

Table 3.1: Addition "mod 12"

as 0, not 12. We can therefore say that addition mod 12 is an isomorphism of clock arithmetic, and that they both represent the same group in some abstract sense. It therefore makes sense to talk about *the* group of addition mod 12 (since there is only one), whether it's represented by numbers, a clock face, or some bizarre variant on the 12 gods of Olympus where Hera + Apollo = Ares.

3.2.3 Rings

But what if our model requires more properties than just addition (as is the case for simple integers, upon which we can perform many more operations)?

Formally speaking, a **ring** is a group with two composition functions instead of just one, normally written as $+$ and $\cdot$. These properties are supposed to capture our intuitions about addition and multiplication, respectively. With respect to $+$, a ring is a normal group, with one additional property: namely, that it "commutes" (exhibits the **commutative** property). This means that for all elements a and b, $a+b=b+a$; it doesn't matter which order they come in. Any group that commutes like this is called an "abelian" group.[3] Not all groups are abelian, as we'll see later. But integers are, and all rings are by definition. The identity element for $+$ is denoted as 0. By convention, the **additive inverse** of x—the number which, when added to another number yields 0—is written $-x$ for a ring; hence $x+-x=0$ as one expects.

The multiplication operation $\cdot$ is associative and has an identity element (written 1). For a group to be a ring, multiplication must also be associative (like addition), and it must follow the standard distributive laws [I.e., $(a+b)c=ac+bc$ and $c(a+b)=ca+cb$]). Normally, a ring must also be commutative with respect to multiplication, but sometimes this is explicitly specified as a "commutative ring."

So a ring is just a group with some other "rules" tacked on. In particular, a ring is a set S and two functions, which we'll describe generally using $\circ$ for the first (as we did with groups) and $\diamond$ for the

[3]Named after the great Norwegian mathematician Niels Henrik Abel (1802–1829). He was so great, in fact, that they also named one of the moon craters after him (Abel Crater). It's on the northwest side of the *Mare Australe*, if you ever find yourself in the neighborhood.

second. The formal properties, then are as follows:

- Both functions are closed over S, which is to say that an element of S $\circ$ an element of S is another element of S, and likewise an element of S $\diamond$ an element of S is another element of S.
- Both functions are associative, which is to say that (x $\circ$ y) $\circ$ z is always equal to x $\circ$ (y $\circ$ z) and likewise for $\diamond$.
- Both functions have identity elements (although those identity elements can be different depending on the ring we're looking at, and normally are).
- The first function ($\circ$) has inverse elements (a value x^{-1} such that $x \circ x^{-1} = x^{-1}$.
- The first function must be commutative, such that $x \circ y = y \circ x$ for all values of x and y.
- Finally, the two functions must be distributive, so that $(x \circ y)z = xz \circ zy)$ and $(x \diamond y)z = xz \diamond zy)$. This is the key property that distinguishes a ring from a group; a ring has two functions, and this property describes how they interact.

As a minor point, we've established that the first function ($\circ$) forms a group. The second function ($\diamond$) is not necessarily a group, because the $\diamond$ function doesn't necessarily have an inverse. The second function is called a **monoid**, which is basically an "almost-group."

Once again, this set of ideas capture the key properties of addition and multiplication as expressed over the integers, and allows generalization of these properties to new domains. As a simple example, the polynomials of a given variable (say, x, which would include elements like $3x^5 + 2x^3 - 7x + 4$) form a ring. We can add two polynomials, subtract them, or multiply them in our normal way; any system we can describe with polynomials can therefore be operated upon as if it were just numbers, provided that we restrict the operations we perform. We can't always divide one polynomial by another, but rings don't require this.

Another example of a ring is the set of integers modulo 12 (as above). This has already been shown to be a group, but it takes little imagination to see that it's also a ring, where multiplication has its normal interpretation. In other words, $3 \cdot 3 = 9$, while $0 + (3 \cdot 5) = 3$; a midnight showing of five three-hour movies will last until 3 o'clock. Again, we can't divide—but that only calls attention to the relationship that is missing and to the peculiarity of the operation.

3.2.4 Fields

To allow for division (the inverse of multiplication) we can create the idea of a "field." A **field** is most simply defined as a commutative ring in which multiplication forms a group over every element except for the additive identity (0). In other words (and this definition assuredly needs them), any element x except for 0 has a unique multiplicative inverse written x^{-1}, with the property that $x \cdot x^{-1} = x^{-1} \cdot x = 1$.

A field is arguably the most important algebraic structure, because it captures most of the arithmetic we remember from elementary school. Note, by the way, that the integers do *not* form a field, since integers generally don't have multiplicative inverses (that are themselves integers). However, fractions do (the inverse—the "reciprocal"—of 2 is just $\frac{1}{2}$, and the inverse of $\frac{2}{3}$ is of course $\frac{3}{2}$); thus we have the ring of integers, but the field of rational numbers, which includes fractions.

Once again, a field is a ring with some additional properties. In particular, a field is a set S and two functions (we'll use $\circ$ and $\diamond$ again) with the following properties:

- Both functions are closed over S.
- Both functions are commutative, associative, and have identity elements in S.
- The first function ($\circ$) has inverse elements.
- Both functions are distributive.

These, of course, are just the properties of a ring, restated. But we now add:

- the second function ($\diamond$) has inverse elements for every element except for the first function's identity element.

As we have seen, the face of a clock—or the natural numbers mod 12—form a ring. They do not, however, form a field. To see why, consider the multiplicative inverse of 2 (2^{-1} or $\frac{1}{2}$). By the

$2 \cdot 0 = 0$	$2 \cdot 1 = 2$	$2 \cdot 2 = 4$	$2 \cdot 3 = 6$
$2 \cdot 4 = 8$	$2 \cdot 5 = 10$	$2 \cdot 6 = 0$	$2 \cdot 7 = 2$
$2 \cdot 8 = 4$	$2 \cdot 9 = 6$	$2 \cdot 10 = 8$	$2 \cdot 11 = 10$

Table 3.2: Searching for 2^{-1} in the ring of integers modulo 12

definition of a field, if this inverse exists, $2 \cdot 2^{-1} = 1$. However, as we can see from table 3.2, no such value exists. There is also no value such that $2 \cdot x = 1$. In fact, there are many such holes in the table; there's no pair that will give you 3, either. What we see instead is that, for example, the expression $2 \cdot x = 8$ has two values for x that satisfy it. Similarly, the $2 \cdot 6 = 0$, even though both 2 and 6 themselves are non-zero. This idea of a "zero divisor"—two nonzero numbers that multiply to give you zero—turns out to be algebraically important. The numbers you learned about in school don't have any zero divisors. The idea violates some of our long-held intuitions about numbers; for example, if in school, we were asked to "solve the equation"

$$x^2 - 5x + 6 = 0 \tag{3.1}$$

you would start by factoring it

$$x^2 - 5x + 6 = (x - 2)(x - 3) = 0 \tag{3.2}$$

and then you would know, that *because the product of two numbers can only be zero if one of the numbers itself is zero*, that the two solutions can only be 2 and 3. This observation and procedure hold true as long as the ring in which you are working has no zero divisors.

$1 \cdot 1 = 1$	$2 \cdot 6 = 12 = 1$	$3 \cdot 4 = 12 = 1$
$5 \cdot 9 = 45 = 1$	$7 \cdot 8 = 56 = 1$	$10 \cdot 10 = 100 = 1$

Table 3.3: Illustration of the inverses of the integers modulo 11

$2 \cdot 0 = 0$	$2 \cdot 1 = 2$	$2 \cdot 2 = 4$	$2 \cdot 3 = 6$
$2 \cdot 4 = 8$	$2 \cdot 5 = 10$	$2 \cdot 6 = 1$	$2 \cdot 7 = 3$
$2 \cdot 8 = 5$	$2 \cdot 9 = 7$	$2 \cdot 10 = 9$	

Table 3.4: 2 is not a zero divisor in the ring of integers modulo 11

By contrast, the ring of integers modulo 11 is a field, as table 3.3 shows. Every element x has a unique inverse. Similarly, table 3.4 shows that 2 is not a zero divisor in this ring.

These two observations are not unrelated. Much of what research mathematicians actually do is make these kind of observations and see if they can be related or explained in terms of each other. As an example, we will offer our very first full-fledged mathematical proof.[4]

Theorem 1. *Every ring with a finite number of elements and no zero divisors is a field.*

Proof. Enumerate the elements of the ring as follows:

$$0, 1, a_1, a_2, a_3, \ldots a_n$$

[4] You may find it helpful to turn back to the definition of a ring and to the last point in the definition of a field.

Now consider any nonzero element x in the ring. For it to be a field, there must be an element x^{-1} such that $x \cdot x^{-1} = 1$.

We first observe that for any two different elements a_i and a_j, the products $x \cdot a_i$ and $x \cdot a_j$ must be different. If they were the same (that is if $x \cdot a_i = x \cdot a_j$) then $0 = x \cdot a_i - x \cdot a_j = x \cdot (a_i - a_j)$. If that were true, then x would have to be a zero divisor, contradicting the assumption that the ring has no zero divisors. We then consider the products

$$x \cdot 1, x \cdot a_1, x \cdot a_2, x \cdot a_3, \ldots x \cdot a_n,$$

Because the a_n are all different (and all different from 1), these products are themselves all different. And because we know the ring has no zero divisors, none of the products are zero. There are therefore exactly as many products as there are non-zero elements of the ring. This implies that these products include every non-zero element $1, a_1, a_2, a_3, \ldots a_n$ (possibly in a different order): one of which, crucially, must be 1. Therefore, for some value a_k, $x \cdot a_k$ equals 1. The only way for that to be true, is for x to have an inverse, which would make the ring a field. Q.E.D. □

Still with us? Good. Because this example shows what the x's really do in algebra. We know something about the structure of a ring, and we know that the ring we're studying has no zero divisors—that's given in the problem statement. We have to show that every element has an inverse; we do this by picking a single element, without caring which one it is, and showing that the element we picked has an inverse. But because the element we picked could have been any element whatsoever, it doesn't matter which element we pick—they all have inverses. The x is therefore just a placeholder for an element of a structure; we may not know anything about that element in particular, but we know enough about the structure in which it is embedded to determine certain properties that it must have. Research in mathematics (and specifically in algebra) is more or less all about finding new properties that are implicit in a given structure. These properties, moreover, help define ways of abstracting the real, physical world in ways that can be modeled and intelligently manipulated.

So now that we have this rather abstract statement in the theorem above, what does it really mean? Among its other implications, it gives us a way to figure out whether or not something is a field. It also gives us (via a corollary we offer without proof) a way of constructing fields if we need them.

Corollary 1. *If p is a prime, the ring of integers modulo p is a field.*[5]

3.3 Further Examples

There are, of course, other algebraic structures that can be studied. The three main ones discussed here form the basis of a lot of applied algebra where we need to represent other parts of the world that can't just be represented as simple numbers.

[5]Informally: to prove this, show that if p is a prime, then there are no zero divisors, since the prime has no divisors other than itself and 1.

3.3.1 Classical Examples

There are a few classical examples of algebraic structure derived from pure mathematics with which you should be familiar. In many cases, these are the type-models from which the concepts such as group and ring were derived.

- $\mathbb{N}$, the natural numbers, are the non-negative counting numbers starting from 0, i.e. (0,1,2,...). These numbers do not form any of the algebraic structures (such as groups) discussed in this chapter. In particular, note that although 0 is an additive identity over this set (e.g. $1+0=1$), and the set is closed under both addition and multiplication (both operations on the set yield other elements within the set), numbers other than 0 do not have additive or multiplicative inverses in this set, since there aren't any negative numbers.

- $\mathbb{Z}$, the integers, are produced by extending the set $\mathbb{N}$ to include negative integers. This forms a ring under the ordinary interpretations of addition and multiplication, but not a field because most numbers don't have multiplicative inverses.

- $\mathbb{Q}$, the rational numbers, are produced by taking ratios of two integers (elements of $\mathbb{Z}$). They form a full-fledged field.

- $\mathbb{R}$, the real numbers, are an extension of the rationals to close it under other operations (for example, taking square roots of positive numbers). It includes not only the rational numbers, but the irrational numbers (such as $\sqrt{2}$), which cannot be expressed as a ratio of two integers. Like $\mathbb{Q}$, they also form a field.

- $\mathbb{I}$, the irrational numbers, are all of the members of $\mathbb{R}$ that are not rational numbers ($\in \mathbb{Q}$). This is specifically not even a group, as the number 0 is not an irrational number, so there is no additive identity element.

- $\mathbb{C}$, the complex numbers, are a field that extends $\mathbb{R}$ to include square roots of negative numbers as well. Unless you already have a good math background, this one is probably not familiar to you, but it's important because you can do things in it that you can't do in $\mathbb{R}$. In $\mathbb{C}$, every polynomial, including apparently unsolvable ones like $x^2 + 1 = 0$, have solutions. This will turn out to be important, for instance, in chapter 6, when we discover that some systems of equations only have solutions if you use complex numbers, and yet these solutions correspond closely to the physical world.

- Imaginary numbers are the counterpart to real numbers in $\mathbb{C}$ although they have no standard symbol by themselves. These are the numbers produced by taking the square root of a negative (real) number. By convention, the square root of -1 is called 'i'; The square root of -4 is therefore 2i, and the square root of -0.25 is 0.5i. Imaginary numbers form a group[6] under addition (2i + 2i = 4i), but are not a ring or a field.

[6] if you accept that $0 = 0i$ is an imaginary number

- $\mathbb{R} - \{0\}$ is an interesting group, produced by taking the field of real numbers and ignoring 0. This forms a group under multiplication, but not a ring.

- $\mathbb{Z}/n\mathbb{Z}$ is the ring of integers "modulo *n*," where *n* is any sensible positive integer. Clock arithmetic, as has already been discussed, is a system equivalent to $\mathbb{Z}/12\mathbb{Z}$. In the case where *n* is prime, this ring is actually a field as discussed above. Since 2, in particular, is prime, the system $\mathbb{Z}/2\mathbb{Z}$ is also a field, and a very important one. If you consider 0 to be equivalent to "false" and 1 to "true," this gives us a representation of logic as an algebraic process. This was also one of George Boole's (1815-1864) great insights after which Boolean algebra and Boolean logic (the same thing, really) are named. Boolean logic/algebra also of course forms the basis for modern computer technology, where different voltage levels represent 0 and 1, respectively.

3.3.2 Vectors

Other applications of algebra incorporate structure in things other than just pure numbers. One of the simplest non-numeric structures is the "vector." A **vector** is simply a ordered collection of arbitrary objects. Often the elements in a vector will themselves be numbers, but they can be almost anything. For example, a set of directions, like the output of a program like MapQuest, is a vector of "turns." Such structures appear in some form in most programming languages as a way of holding ordered collections of objects for later manipulation. Computer scientists sometimes call this kind of structure a "container."

Vectors are often used to represent positions. For example, positions on a globe (latitude and longitude) can be represented by a two-place vector. London, for example, is at latitude 52N, longitude 0W, while New York City is at 41N, 74W. Letting positive numbers denote north latitude and west longitude, we could equally describe the cities' locations as (52,0) and (41,74) respectively. Amsterdam is at (52,-4), Ankara is at (42,-32), and Perth is at (-32,-116). Alternatively, one can use the same notation to describe trips. A trip five degrees due west would be represented as (0,5). One can then do addition and subtraction on the individual components to determine differences between locations. Traveling five degrees west (0,5), and three degrees north (3,0) would result in an overall trip of (3,5). The difference between New York and Amsterdam is (-11, 78), meaning that to fly from Amsterdam to New York, you would need to head in an overall direction just south of west until you had covered 11 degrees of latitude and 78 of longitude.

Such position vectors are structured as a group. Any two elements can be added by independently adding the components of the vector; the sum of two vectors represents the overall position you get by making two trips in succession. The inverse of a vector can be obtained by taking the inverse of the individual components; the "identity" vector (0,0) represents traveling no distance whatsoever.

Of course, vectors can be used in other ways. The Manhattan street layout lends itself well to a two-dimensional vector, as does a chessboard. The vector (-2,1) could be used to represent one of the possible moves of the knight. A bishop can move only diagonally, which we can represent by the abstract vector scheme (x,x) or $(x,-x)$ (thus capturing the idea that the change in rank must be equal to the change in file). White pawns, of course, have only (1,0) as their move except when they capture using (1,1) or (1,-1) or when they move (2,0) initially.

Vectors can have as many places as one needs; moving around in three dimensions as a rocket or a plane might can be captured using a three-place vector (x,y,z). Alternatively, the elements of a vector may not represent physical dimensions, but different abstract elements of an idea. For example, the inventory of an ice cream shop might be represented by a fifty-element vector, with the numbers representing the number of gallons of chocolate, vanilla, strawberry, coffee, etc. in stock. Adding vectors would change the stock, as people buy ice cream and new supplies arrive.

It's fairly easy to formalize these concepts with a strict mathematical definition. An n-place vector a is an ordered collection of n objects, written $(a_1, a_2, a_3, \ldots a_n)$. Vectors are added (using any addition operation valid for the underlying objects) by adding together corresponding elements:

$$(a_1, a_2, \ldots, a_n) + (b_1, b_2, \ldots, b_n) = (a_1 + b_1, a_2 + b_2, \ldots, a_n + b_n) \tag{3.3}$$

This operation is closed over the set of vectors and is obviously associative if the underlying element-operation is.

In this definition, the identity vector can be seen to be an n-place vector of all zeros : $(0,0,\ldots,0)$. For any vector a, the inverse is a vector containing the individual element inverses:

$$-a = (-a_1, -a_2, -a_3, \ldots, -a_n) \tag{3.4}$$

(Note that $a + -a = 0$ as required.) Therefore, the set of n-place vectors forms a group.

We can also define the operation of **scalar multiplication**—multiplying a vector by a constant (a number)—as the result of multiplying each element by that constant:[7]

$$k \cdot (a_1, a_2, \ldots, a_n) = (k \cdot a_1, k \cdot a_2, \ldots, k \cdot a_n) \tag{3.5}$$

Thus 3 · (4,5,6) = (4,5,6) + (4,5,6) + (4,5,6) = (12,15,18) as we expect, and multiplying a vector by -1 yields its inverse.

So why are vectors a group and not a ring or a field? Although the idea of adding vectors makes intuitive sense and is well-defined, it doesn't make quite as much sense to talk about multiplying New York by London. Although scalar multiplication, above, is well-defined, you can't use this operation to multiply one vector by another in all circumstances, as a ring would require. The standard definition of multiplying one vector by another does not fulfill the definition of a ring.

In particular, vector multiplication (multiplying one vector by another) is usually defined as the sum of the individual element products:

$$(a_1, a_2, \ldots, a_n) \cdot (b_1, b_2, \ldots, b_n) = a_1b_1 + a_2b_2 + \ldots + a_nb_n \tag{3.6}$$

[7] In the context of vectors and matrices, an ordinary number is usually called a **scalar**. Multiplying a vector or matrix by a scalar produces a product for every element (essentially re-"scaling" it, hence the name). For example, multiplying the vector (1,2,3,4,5) by 3 yields a rescaled vector (3,6,9,12,15).

$$\begin{bmatrix} a_{11} & a_{12} & \cdots & a_{1n} \\ a_{21} & a_{22} & \cdots & a_{2n} \\ \vdots & \vdots & & \vdots \\ a_{m1} & a_{m2} & \cdots & a_{mn} \end{bmatrix}$$

Figure 3.4: A sample matrix

In this formulation, the product of a vector and a vector is not a vector, but just a single number, and hence vectors are not closed under multiplication. Of course, nothing stops someone from defining a different, non-standard, multiplication algorithm for vector-on-vector multiplication, but so far there has not been any real need or application.

3.3.3 Matrices

Instead, most applications that require multiplication of vectors tend to be re-designed to use matrices instead. While a vector is a one-dimensional container, a **matrix** is a two-dimensional container. Technically, for positive integers m and n, an $m \times n$ matrix is a collection of mn numbers or other elements arranged in a rectangular array with m rows and n columns, as in figure 3.4.

For example, $\begin{bmatrix} 8 & 1 & 6 \\ 3 & 5 & 7 \\ 4 & 9 & 2 \end{bmatrix}$ is a 3×3 matrix. Because m and n are equal, this is also sometimes referred to as a **square matrix**.

The definition of a matrix therefore includes vectors, since an n-

place vector is equivalent to a $1 \times n$ (row) matrix or an $n \times 1$ (column) matrix. Matrix addition is defined when two matrices are the same size as the sum of corresponding matrix elements; this again includes what we have previously defined as vector addition. Scalar multiplication can also be defined (as it is with vectors) as the product of some number (the **scalar value**) and each element of the matrix.

Are matrices, in general, a group? No. In general, one can't add matrices of two different sizes, such as a 2×3 and a 4×5 matrix. But matrices of the same size form a group, where the additive identity is the matrix of all zeros, and the additive inverse is simply -1 times the original matrix. Matrix multiplication *has* been defined so that square matrices of a given size form a useful ring. We'll explain that in a bit, but it will help to show how matrices are useful representations first. To keep our eyes on the loftier matters of abstract algebra, we'll use a couple of easy-to-understand examples (that are highly reminiscent of high-school algebra problems).

To begin, let's consider the problems of a soft drink factory that makes several different types of drinks. Each drink type needs different amounts and types of raw materials, and of course, each material has its own properties such as cost.

A matrix can be used to hold the numbers relating each type to each material. For example, table 3.5 describes a set of materials (water, sugar, flavorings, and bottles) for various types of drinks. In this case, each drink takes exactly one bottle, but the small drinks take less water and proportionately less sugar. The numbers in this table can be captured in a matrix more or less exactly as written.

Suppose the corporation wants to calculate the material costs of

Drink	Water (litres)	Sugar (grams)	Flavor (grams)	Bottles (each)
Large Fizzy-Pop	1	100	10	1
Small Fizzy-Pop	0.5	50	5	1
Large Slim-Pop	1	0	10	1
Small No-tayst	0.5	50	0	1

Table 3.5: Some thoroughly disgusting drink components

Item (unit)	Cost/unit (cents)
Water (per liter)	10
Sugar (per gram)	8
Flavor (per gram)	100
Bottles (each)	10

Table 3.6: Costs for some thoroughly disgusting drink components

each type of drink. In general, the expression for cost per drink is given by the following equation

$$cost/drink = units/drink \cdot cost/unit$$

The total cost of the drink is of course the sum of the individual material costs.

We express the costs of the individual components as a column vector (the reason for using columns will be apparent soon), as in table 3.6. (For example, sugar costs $0.08 per gram.)

The cost of any given drink is therefore the sum of the products of the amount of the component and the cost of the component. In general, for an $l \times m$ matrix and a $m \times n$ matrix, the matrix product of the two matrices is a third, $l \times n$ matrix. Each entry of this product matrix is defined as the sum of the products of the corresponding row of the first matrix and column of the second.

Here's the equation we'll be using for this example:

$$\begin{bmatrix} 1 & 100 & 10 & 1 \\ 0.5 & 50 & 5 & 1 \\ 1 & 0 & 10 & 1 \\ 0.5 & 50 & 0 & 1 \end{bmatrix} \cdot \begin{bmatrix} 10 \\ 8 \\ 100 \\ 10 \end{bmatrix} = \begin{bmatrix} a \\ b \\ c \\ d \end{bmatrix}$$

In this example, the value of a, the first row of the first (and only) column of the answer will be the sum of the corresponding elements of the first row with the first (and only) column.

$$a = (1 \cdot 10) + (100 \cdot 8) + (10 \cdot 100) + (1 \cdot 10) = 1820$$

which is of course the cost of a large Fizzy-Pop. A small Fizzy-Pop costs b, the product of the second row and (only) column.

$$b = (0.5 \cdot 10) + (50 \cdot 8) + (5 \cdot 100) + (1 \cdot 10) = 915$$

and so on for c, d and any other drinks we want to manufacture.

One way to think about this process is to visualize the number 7, with its horizontal bar to the left of the vertical downstroke, showing that you take the row of the first argument and multiply it with the column of the second—the sort of thing that could be automated on a computer with great ease, even over terrifically large sets of data. This technique has such remarkable versatility, in fact, that can be applied to a truly astounding variety of systems and processes. In the nineteenth century, mathematicians developed this notation as a shorthand for solving systems of so-called "linear equations."

A **linear equation** is simply an equation like

$$Ax_1 + Bx_2 + Cx_3 + \ldots + Yx_n = Z$$

where a final result is the sum of a number of products of independent variables and their corresponding coefficients. These systems arise in all sorts of contexts. The soft drink example is a good example, where the final prices are the sum of several subtotals determined by unit prices and amounts. Demographically, the total population of a given religion is the sum of the percentage of adherents in each district, times the population of that district. And so forth.

Conversely, you can use systems of equations like this to solve other problems, where you know overall properties but not individual ones. By inspecting the cost structure for the soft drink company, one might be able to infer the price it pays for sugar, perhaps in the interest of some sort of dumping or anti-trust suit. But to do that takes a little more mathematical machinery.

Another high-school style mathematical puzzle: A pet store sells kittens at $80 and puppies at $100. If I bought 8 pets for $720, how many of each did I buy? The problem implicitly lends itself to the following structure as two linear equations. If we let k be the number of kittens I bought, and p the number of puppies, then we have the following relationships:

$$80k + 100p = 720 \quad \text{(The total amount I spent is \$720.)}$$
$$k + p = 8 \quad \text{(The number of animals I bought is 8.)}$$

which can be represented as the following set of matrices:

$$\begin{bmatrix} 80 & 100 \\ 1 & 1 \end{bmatrix} \cdot \begin{bmatrix} k \\ p \end{bmatrix} = \begin{bmatrix} 720 \\ 8 \end{bmatrix}$$

If we call the $\begin{bmatrix} 80 & 100 \\ 1 & 1 \end{bmatrix}$ matrix A, call the $\begin{bmatrix} k \\ p \end{bmatrix}$ matrix B, and the $\begin{bmatrix} 720 \\ 8 \end{bmatrix}$ matrix C, then it actually looks like a rather simple problem. It's just $A \cdot B = C$, right?

Actually, it really is just that simple. If we could find another matrix A^{-1} (the multiplicative inverse of A), we could just multiply both sides by A^{-1} and solve for B.

$$\begin{aligned} A \cdot B &= C \\ A^{-1} \cdot A \cdot B &= A^{-1} \cdot C \\ (A^{-1} \cdot A) \cdot B &= A^{-1} \cdot C \\ 1 \cdot B &= A^{-1} \cdot C \\ B &= A^{-1} \cdot C \end{aligned}$$

So B is just the product (which we know how to find) of A^{-1} and C.

The key algebraic question, though, is "do matrices have inverses?"

In order to have an inverse, there must first be a multiplicative identity. Fortunately, such an identity is not difficult to find. The "identity matrix" is simply a square matrix with ones along the main diagonal (upper left to lower right).[8] It's fairly easy to confirm, for example that

[8]This technically means that there are really several different identity matrices,

$$\begin{bmatrix} 1 & 0 & 0 & 0 \\ 0 & 1 & 0 & 0 \\ 0 & 0 & 1 & 0 \\ 0 & 0 & 0 & 1 \end{bmatrix} \cdot \begin{bmatrix} w \\ x \\ y \\ z \end{bmatrix} = \begin{bmatrix} w \\ x \\ y \\ z \end{bmatrix}$$

no matter what values you pick for w, x, y, z. To find an inverse matrix A^{-1}, one simply needs to find a matrix that yields the 2×2 identity matrix. This process—finding the inverse of a matrix—can be mathematically and computationally challenging. But although the math can be difficult, the concept really isn't.

Here, we'll simply cut to the chase. For this particular set of equations, A^{-1} does exist and its value is $\begin{bmatrix} -\frac{1}{20} & 5 \\ \frac{1}{20} & -4 \end{bmatrix}$. Plugging this value in and doing the multiplication, we see that $\begin{bmatrix} k \\ p \end{bmatrix}$ is $\begin{bmatrix} 4 \\ 4 \end{bmatrix}$, or equivalently, that I bought 4 kittens and 4 puppies.

So does this mean that matrices form a field? Not quite, but they do form a ring. The main problem is that while many matrices have inverses, not all do. The matrices that do not have inverses correspond to systems of equations that don't have solutions. But because systems that can be solved correspond to matrices that do have inverses, we can use normal rules of arithmetic to solve complex problems involving not just single numbers, but huge groups of them.

---

since the 4×4 identity matrix is different from the 3×3. This isn't a big problem as long as you understand the concept.

3.3.4 Permutations

Another important example of group structure appears when we rearrange elements in a list. For example, the numbers 1 through 4 in their usual order are (1,2,3,4), but can be rearranged by swapping the first two elements into (2,1,3,4).[9] Things other than numbers can also be permuted, as in the earlier combinatorics problems.

Permutations form a rather unusual group, in that they are not commutative, and doing the same permutations in a different order can yield different results. For example, consider the following two permutations: (2, 1, 3, 4) and (1, 3, 2, 4), essentially swapping the first two and the middle two elements, respectively. Notice that the results of these two permutations are different when applied in a different order to the same set:

$$
\begin{array}{rl}
Alpha, Bravo, Charlie, Delta & Alpha, Bravo, Charlie, Delta \\
Bravo \Leftrightarrow Alpha, Charlie, Delta & Alpha, Charlie \Leftrightarrow Bravo, Delta \\
Bravo, Charlie \Leftrightarrow Alpha, Delta & Charlie \Leftrightarrow Alpha, Bravo, Delta
\end{array}
$$

Permutations are perhaps more important in theoretical mathematics than they are in practice; there are few direct applications. However, one major application is of crucial importance—process control, or "in what order do we want to do the necessary steps?" In many applications in science, technology, and business, doing

[9]You'll recall from the previous chapter that a permutation is simply a possible ordering of a collection of objects; there are 22 other other ways to rearrange these numbers.

steps in the most efficient way possible is crucial to getting the best possible results as fast as possible. Permutations provide an abstraction for studying this problem. Permutations are also key to many problems in cryptography (a matter we will touch upon in the final section).

3.3.5 Topological Transformations

Any problem involving data visualization must take into account the idea of spacial transformations. For example, a penny that appears circular when seen from the top may appear oval or even as a straight line from an oblique view. Depending upon the angle you look at it, a square cube of sugar might have a hexagonal outline.[10] **Topology**, a rather advanced branch of mathematics, is the study of the sort of distortions you can get when you continuously transform an object, moving and perhaps bending it, but not breaking it.

One important transformation in graphical manipulation is the idea of an "affine transformation." As an example, consider the transformations of the sailboat in figure 3.5. It's been moved (translated), rotated, re-scaled, and generally messed with.

An **affine transformation** is a special type of transformation that does not distort straight lines (i.e. lines remain straight after the

[10] The French proto-absurdist author Alfred Jarry once wrote, "Universal assent is already a quite miraculous and incomprehensible prejudice. Why should anyone claim that the shape of a watch is round—a manifestly false proposition—since it appears in profile as a narrow rectangular construction, elliptic on three sides; and why the devil should one only have noticed its shape at the moment of looking at the time?" [7]

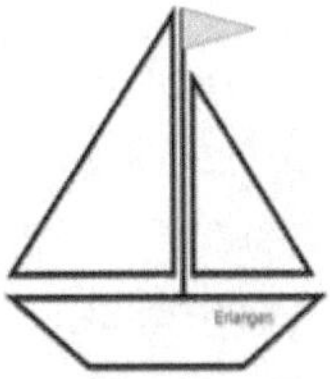

Original image

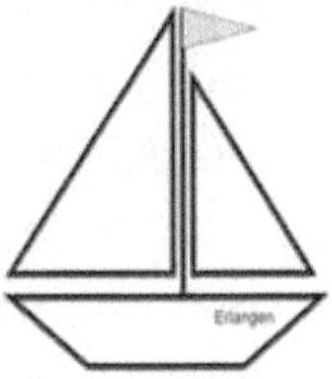

Unchanged image moved right

Image rotated

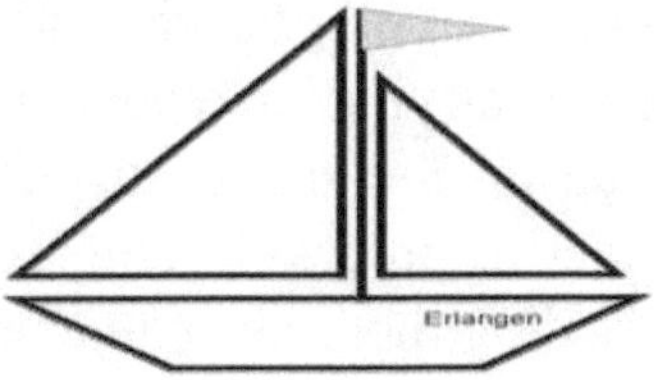

Image rescaled horizontally

Figure 3.5: Simple affine transformations

transformation; the transformation of a square into a circle would not be affine). All of the above are examples of affine transformations. As you might expect from the previous discussion, affine transformations form another algebraic group. In fact, they form a group that can be represented and manipulated using the vector and matrix representations developed earlier.

Any point or pixel on a screen can of course be represented by a vector in two dimensions, and any point in a 3D model can be represented by a vector in three. Since lines are straight (and remain straight under these transformations), one can represent a more complex shape as a collection of points/vectors—a line by its endpoints, a triangle by its three corners, and a heptakaidecahedron by ...whatever happens to apply in that case. The affine transformations in this framework represent transformations that can be employed in the visualization of such structures, including moving objects in space, magnifying or reducing them, and projecting them onto a two-dimensional plane for display. These operations become a part of the rendering process in a high-end graphics display system, and by understanding them, we can come to a better understanding of how to display or manipulate data and visualizations.

These operations are relatively easy to incorporate using the vector/matrix framework. Each point can be represented by a column vector, so a polygon of m points can be represented by a $2 \times m$ matrix. (For shapes in three dimensions, use $3 \times m$). For explanatory purposes, we will use the matrix

$$A = \begin{bmatrix} 1 & 1 & 0 \\ 0 & 1 & 1 \end{bmatrix}$$

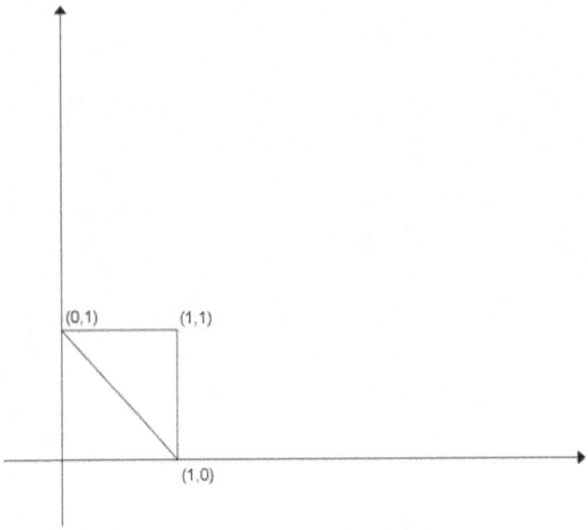

Figure 3.6: Simple figure for affine transformations

a simple right triangle located near but not on the origin. (See figure 3.6.)

Affine transformations can be performed by elementary matrix operations such as addition and multiplication. For example, the point $\begin{bmatrix} a \\ b \end{bmatrix}$ can be translated (moved) by adding the desired displacement directly to the coordinate. As an example, adding the vector

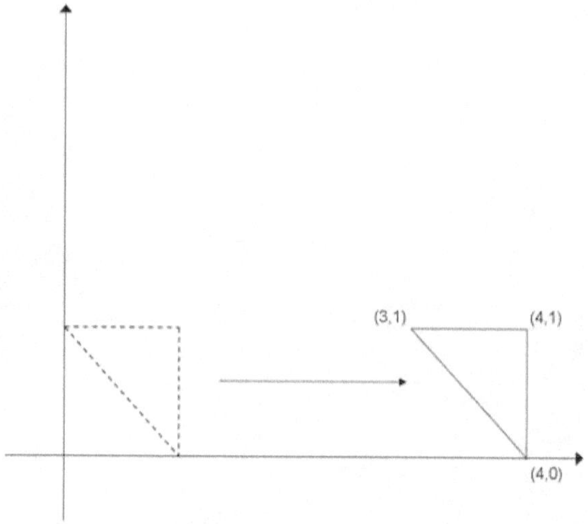

Figure 3.7: Simple figure moved (formally, translated)

$\begin{bmatrix} 3 \\ 0 \end{bmatrix}$ directly to each point of the triangle creates the new matrix

$$\begin{bmatrix} 4 & 4 & 3 \\ 0 & 1 & 1 \end{bmatrix}$$

moving the entire triangle to the right as in figure 3.7.

We can then use scalar multiplication to resize the triangle. For example, $3 \cdot A$ yields a triangle three times as large. (Figure 3.8) More generally, we can rescale the triangle independently in different axes

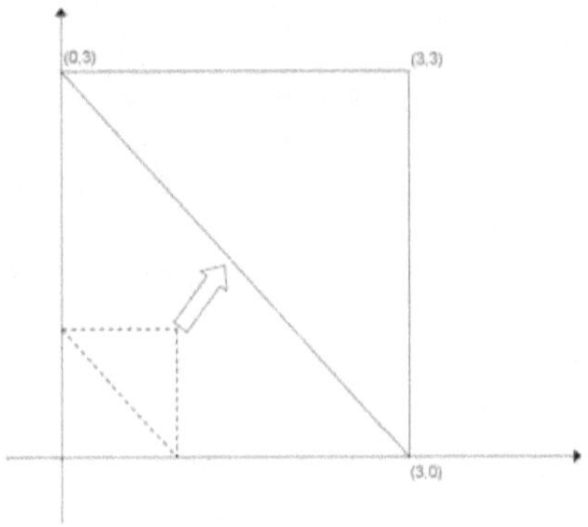

Figure 3.8: Simple figure enlarged three times

using an appropriate matrix multiplication. A tall, skinny triangle (narrowing the triangle in the first coordinate, enlarging it in the second, as in figure 3.9) can be be obtained by the following,

$$A' = \begin{bmatrix} 0.5 & 0 \\ 0 & 3 \end{bmatrix} \cdot \begin{bmatrix} 1 & 1 & 0 \\ 0 & 1 & 1 \end{bmatrix} = \begin{bmatrix} 0.5 & 0.5 & 0 \\ 0 & 3 & 3 \end{bmatrix}$$

In general, multiplying by $\begin{bmatrix} \alpha & 0 \\ 0 & \beta \end{bmatrix}$ will rescale the image along the first dimension by α, enlarging the image if $\alpha > 1$, reducing it if

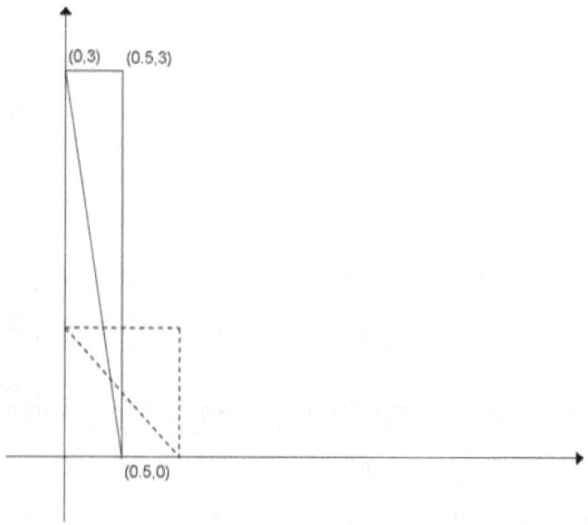

Figure 3.9: Simple figure rescaled vertically and horizontally

less, and will rescale the image along the second by β in the same way.

What happens if α or β is exactly 1? Well, rescaling an image by one will produce a new image of exactly the same size (which suggests, correctly, that $\begin{bmatrix} 1 & 0 \\ 0 & 1 \end{bmatrix}$ is an identity element). More interestingly, what happens if α or β is zero? In this case, it will reduce the image in that direction infinitely far, down to a single line.

This produces the transformation called a **projection**, reducing one of the dimensions involved in the original image by one. For example, multiplying an image by $\begin{bmatrix} 1 & 0 \\ 0 & 0 \end{bmatrix}$ will preserve the data in the first column unchanged (because α is one), but make the second column of data all zeros (because β) is zero. This isn't very interesting when we're dealing with only two-dimensional data, but can be key to visualizing three (or more) dimensional data, by projecting it down to a two-dimensional set that can be easily shown on a printed page or computer screen. A traditional map, for example, projects the three-dimensional world down to a 2D image, so that the top of the Empire State building and the surface of the Hudson River are both at the same physical level on the flat sheet of paper.

Rotation about the origin can be obtained by simple matrix multiplication using trigonometric functions. As in figure 3.10, the new point (x', y') is related to the old one (x, y) after a rotation through the angle θ by these equations

$$x' = x\cos\theta - y\sin\theta \qquad (3.7)$$

$$y' = x\sin\theta + y\cos\theta \qquad (3.8)$$

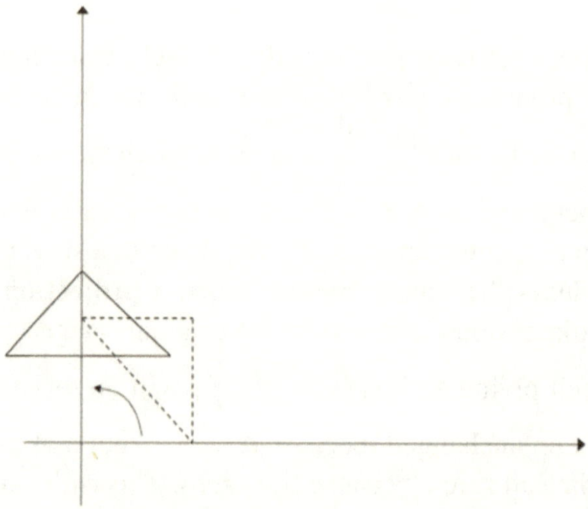

Figure 3.10: Simple figure rotated around origin

which can be achieved in matrix form by

$$\begin{bmatrix} x' \\ y' \end{bmatrix} = \begin{bmatrix} \cos\theta & -\sin\theta \\ \sin\theta & \cos\theta \end{bmatrix} \cdot \begin{bmatrix} x \\ y \end{bmatrix} \tag{3.9}$$

(If you don't believe this, a few minutes of work with trigonometric functions and polar coordinates should convince you. Or you can just take our word for it.)

However, what if we want to rotate about some point other than just the origin? This is an example of where the group (actually,

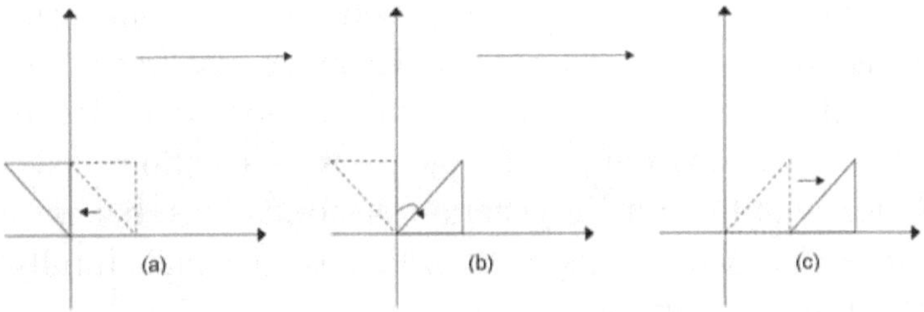

Figure 3.11: Rotation around arbitrary point as three step process

ring) structure of matrix operations is helpful. Because ring operations can be composed, we are at liberty to build complex operations such as general rotations from the simpler ones we have shown. In particular, we can rotate an image around an arbitrary point by a combination of three steps, as shown in figure 3.11. First, translate the image so that the center of rotation is at the origin [(a)]. Second, rotate about the origin [(b)], and third, translate the rotated image back to its original location [(c)]. This can be expressed mathematically using the following equation—here A represents the original image, d represents the location of the point of rotation, and R the matrix to do the rotation. A' is, of course, the final (rotated) image.

$$A' = (R \cdot (A - d)) + d \tag{3.10}$$

There are other forms of affine transformations that have not been discussed here for reasons of brevity: shears and reflections among others. Our discussion has been limited to two dimensions for the same reason, but images or representations in three dimensions (or higher) are obvious generalizations. There are also transformations that are not affine – transforming a (round) globe to a (flat) map or vice versa is a good example. Topology, as we mentioned, is deeply concerned with these more general transformations, but we would need more mathematics, especially calculus, to explain it fully. And that's still two chapters away.

The central point remains that by taking advantage of the structure of the kind of transformations useful and interesting in real life, capturing that structure in the formal definition of affine transformations, and then representing that in the ring structure of matrices, you can describe almost arbitrarily complex transformations in terms of a few simple and easily understandable primitives. The language of mathematical structure—algebra—gives us a handle on the shapes of the world.

3.3.6 Cryptology

A final important use of the structures of algebra can be found in the field of cryptology, since algebra forms the foundation for most modern cryptographic algorithms. Unfortunately, the people who really understand the algebra and algorithms in detail aren't allowed to talk about it. However, some historical examples can serve to illustrate the relationship between codes, ciphers, and mathematical structure.

And really, we don't have to go very far back, because this relationship was only recently discovered. Prior to about 1930, codemaking and codebreaking was dominated, not by mathematicians and computer programmers, but by linguists and literature scholars who were familiar with the structure of language. Cryptography provides a way to hide the structure of language behind a mathematical structure, but understanding the structure of language is still key.

In particular, note that "coding" in general, forms a group. Any process that you use to muddle letters and words up can be repeated or composed with any other process, including the "null" process of not doing anything. This null cipher, in which you cleverly replace 'a' by 'a', 'b' by 'b', 'c' by 'c', and so forth, is obviously the identity element in this group. Furthermore, to be useful, a code or cipher must be invertible, so that you can decode it and figure out what the original message was. These are, of course, the defining properties of groups.

Within this large group are several subgroups that can be individually addressed. For example, one classical cipher is the transposition cypher, where the letters in a message are unchanged but moved about. One common method for doing this involves writing the message into a set of ten or so columns and then writing the columns from top to bottom. The message "Only mad dogs and Englishmen go out in the noonday sun" is placed in columns as in the diagram

O	N	L	Y	M	A	D	D	O	G
S	A	N	D	E	N	G	L	I	S
H	M	E	N	G	O	O	U	T	I
N	T	H	E	N	O	O	N	D	A
Y	S	U	N						

and becomes the encoded message OSHNY NAMTS LNEHU YDNEN MEGN ANOO DGOO DLUN OITD GSIA. This can be a surprisingly strong code; using this code twice (a so-called double transposition cipher) was still used as recently as the Second World War.

Structurally, it should be obvious that this is just a type of permutation, and that to solve this type of cipher it is necessary “only” to find the inverse permutation to put the letters back in their proper order. Furthermore, the muddling introduced by the algebraic permutation structure influences only the position of the letters, but not their identity. This is a point to which we shall return.

Another common cipher (and interesting subgroup) is the substitution cipher, typified by a newspaper cryptogram (see figure 3.12).[11] In this type of cipher, each letter is replaced by a different letter; for example “a”’ might become “m”, “b” “i” and so forth. This again can be shown to be a type of permutation, this time a permutation of the letters of the alphabet:

[11] The solution is provided at the back of the chapter.

Cgsbdm ctn hddf mdtodn egk mgbd cgqkm lf mlsdfud wloc clm sgfr, oclf htup uqkvdn gvdk t ucdbluts vdmmds lf wcluc cd wtm hkdwlfr t itkoluqstksy btsgngkgqm ikgnquo. Clm cdtn wtm mqfp qigf clm hkdtmo, tfn cd sggpdn ekgb by iglfo ge vldw slpd t moktfrd, stfp hlkn, wloc nqss rkty isqbtrd tfn t hstup ogi-pfgo.

Figure 3.12: Sample cryptogram #1

Plaintext	A	B	C	D	E	F	G	H	I	J	K	L	M
Ciphertext	M	I	B	X	F	O	D	V	A	K	T	R	N

Plaintext	N	O	P	Q	R	S	T	U	V	W	X	Y	Z
Cyphertext	Z	S	G	P	H	Q	C	E	Y	U	J	L	W

Again, solving this cipher requires only that the cryptographer recover the inverse permutation and apply it. Furthermore, as with the transposition cipher, the structure of the cipher hides some, but not all, of the structure of the underlying message.

In this case, for example, part of the structure of English (or any language) includes word structure. In figure 3.12, word boundaries have not been hidden, so the cryptographer can look at the short words and guess what they or their component letters are. We know, for example, that the only two common one letter words are “a” and “I,” so any ciphertext letter in isolation (in this case, the letter “t”) is probably one of those two. (Hint: in this case it’s an ‘a.’)

Even when word boundaries are hidden, as in figure 3.13, the structure of English may still come through via the idea of commonness, or more technically, frequency. Not all letters occur with

LFOCD IKDMD FOUTM DLFND DNLFT SSUTM DMGEM DUKDO WKLOL FROCD ELKMO JQDMO LGFKD RTKNM OCDST FRQTR DGEOC DULIC DKEGK OCDIK LFULI SD-MGE MGSQO LGFMG ETKDM IDULT SSYTM OCDBG KDMLB ISDUL ICDKM TKDUG FUDKF DNNDI DFNQI GFTFN TKDVT KLDNH YOCDR DFLQM GEOCD ITKOL UQSTK LNLGB LFRDF DKTSO CDKDL MFGTS ODKFT OLVDH QODXI DKLBD FONLK DUODN HYIKG HTHLS LOLDM GEDVD KYOGF RQDPF GWFOG CLBWC GTOOD BIOMO CDMGS QOLGF QFOLS OCDOK QDGFD HDTOO TLFDN HQOWL OCOCD ULICD KFGWH DEGKD QMTSS NLEEL UQSOY WTMKD BGVDN HYOCD MLRFT OQKDO CDIQF QIGFO CDWGK NPLNN LMTII KDULT HSDLF FGGOC DKSTF RQTRD OCTFO CDDFR SLMCH QOEGK OCLMU GFMLN DKTOL GFLMC GQSNC TVDHD RQFBY TOODB IOMWL OCOCD MITFL MCTFN EKDFU CTMOC DOGFR QDMLF WCLUC TMDUK DOGEO CLMPL FNWGQ SNBGM OFTOQ KTSSY CTVDH DDFWK LOODF HYTIL KTODG EOCDM ITFLM CBTLF TMLOW TMLTM MQBDN OCDUK YIOGR KTICO GHDDF RSLMC

Figure 3.13: Sample cryptogram #2

equal frequency; the letter "e" usually forms up to 21% or more of the letters in a document, while "q" and "z" are $\frac{1}{2}$% or less. Knowing this, we can guess that the common letters in the second cryptogram correspond to high-frequency English letters such as "e," "t," "a," "o," or "n," and start our decoding appropriately. Other structures—for example, the letter pair "th" is very frequent, but "ht" is almost unseen—can further help guide us, precisely because this information is not masked by the encryption process.

A better code—and of course, all "modern" systems are much better than these—would hide as much regularity as possible precisely to prevent this. But even here, knowledge of the underlying algebraic structure can help guide attempts to solve even modern, difficult codes.

A well-known example of this is the "Enigma" machine used by the Germans during the Second World War to encrypt communications, including those meant for their U-boats and army commanders. Solving this cipher was obviously of critical importance to the countries facing Germany. It was actually Poland, however, that produced the first algebraic attack on the Enigma and helped create the first generation of mathematical cryptologists.

The Enigma is an example of an electromechanical cipher machine of the sort that were popular in the early twentieth century. In appearance, it looked rather like a typewriter with a set of lights attached. The user would set a "key" through a set of wires, plugs, and rotating wheels, then press the keys for the message he wanted to send; the lights would light up corresponding to the ciphertext for each letter. Decryption operated in exactly the same way.

The basic operation of an Enigma machine is shown in figure 3.14.

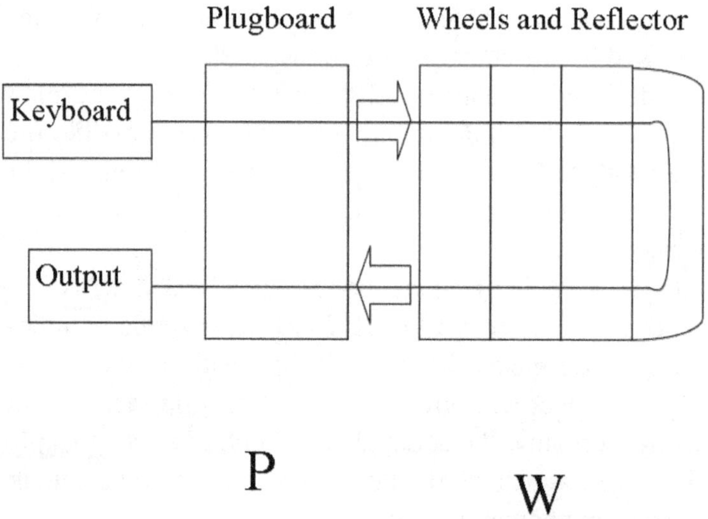

Figure 3.14: Simplified schematic of three-rotor Enigma

Pressing the key would send an electrical signal through a set of plugs in a plugboard, then through a set of rotating wheels, then back through the plugboard in reverse until it lighted a lamp. The variable connections created by the plugboard did a very effective job of masking the individual letters used in the message (like a newspaper cryptogram), but did not hide the frequency or structural properties. The rotating wheels, however, ensured that each letter of the message was encrypted like a separate newspaper cryptogram, essentially masking these self-same frequency properties. The combined effect was, to say the least, formidable.

Marion Rejewski (1905–1980), a Polish mathematician, made history precisely by recognizing the algebraic structure inherent in this system. Either of the two parts, the plugboard or the rotating wheels, could be solved relatively quickly and easily, because of the information they did not mask. Could these two parts be separated in practice?

In particular, one can formalize the operation of an Enigma machine as follows: If we let M denote the message, C the encoded message, P the code created by the plugboard, and W the code created by the rotating wheels, then the full operation can be described as

$$C = M \circ P \circ W \circ P^{-1}$$

representing passing the message M successively through the plugboard, the wheels, and then in reverse (P^{-1}) through the plugboard.

Knowing that the plugboard P did not disturb sequence-dependent relationships, he asked himself whether there were any unusual sequence-dependent relationships that could be inferred from looking at C.

These relationships could only be properties derived from W, and hence could be used to identify W independent of the other cryptographic aspects. His attempt, of course, was successful. By identifying W, he was able to then solve separately the not-too-difficult plugboard cipher, and thus reveal the secrets of the German military.

This incident in some regards marks the turning point in the history of cryptography; today cryptography is largely dominated by mathematicians, precisely because of their knowledge of mathematical structure. The RSA algorithm that keeps much of the Internet secure, for example, is a simple expression of a few number-theoretic properties of the group structure of $\mathbb{Z}/n\mathbb{Z}$. One can't help but wonder what advances might be possible by examining these methods in light of linguistic and literary structure, appropriately modeled as mathematical structure.

3.4 Answers to chapter puzzles

3.4.1 Sudoku

The solution (in numbers) to the sudoku puzzle is:

4	8	6	5	1	7	3	9	2
9	1	5	2	4	3	8	6	7
7	3	2	6	9	8	5	4	1
3	5	7	4	2	6	9	1	8
1	2	9	8	7	5	6	3	4
6	4	8	9	3	1	2	7	5
8	6	1	3	5	4	7	2	9
2	7	3	1	8	9	4	5	6
5	9	4	7	6	2	1	8	3

3.4.2 Cryptograms

Both cryptograms were created using the key "THUNDERCLAPS-BFGIJKMOQVWXYZ", i.e A becomes T, B becomes H, C becomes U, and so forth. The first cryptogram is the opening paragraph of one of the classics of cryptographic fiction, "The Adventure of the Dancing Men," by Conan Doyle:

> Holmes had been seated for some hours in silence with his long, thin back curved over a chemical vessel in which he was brewing a particularly malodorous product. His head was sunk upon his breast, and he looked from my point of view like a strange, lank bird, with dull gray plumage and a black top-knot.

The second is an explanatory paragraph from another such classic, "The Gold Bug," by Poe.

"In the present case - indeed in all cases of secret writing - the first question regards the language of the cipher; for the principles of solution, so far, especially, as the more simple ciphers are concerned, depend upon, and are varied by, the genius of the particular idiom. In general, there is no alternative but experiment (directed by probabilities) of every tongue known to him who attempts the solution, until the true one be attained. But, with the cipher now before us, all difficulty was removed by the signature. The pun upon the word 'Kidd' is appreciable in no other language than the English. But for this consideration I should have begun my attempts with the Spanish and French, as the tongues in which a secret of this kind would most naturally have been written by a pirate of the Spanish main. As it was, I assumed the cryptograph to be English."

4 Probability and Statistics

On November 9th, 1965, over thirty-million people lost power across an area that extended from New York and New Jersey, through New England, and into parts of southern Ontario—an area comprising some 80,000 square miles. On November 10th, *The New York Times* ran the headline:

> POWER FAILURE SNARLS NORTHEAST; 800,000 ARE CAUGHT IN SUBWAYS HERE; AUTOS TIED UP, CITY GROPES IN DARK.

There have been power outages of similar magnitude since, but for many New Yorkers, it still stands as The Great Blackout.

Part of its fame undoubtedly arises from a headline the *Times* ran exactly nine months (to the day) later, on August 10th, 1966:

> BIRTHS UP NINE MONTHS AFTER THE BLACK-OUT.

The story that followed made the case plain:

> A sharp increase in births has been reported by several large hospitals here, 9 months after the 1965 blackout.
>
> Mount Sinai Hospital, which averages 11 births daily, had 18 births on Monday. This was a record for the hospital; its previous one-day high was 18. At Bellevue there were 29 new babies in the nursery yesterday, compared with 11 a week ago and an average of 20.

The sharp rise seemed to be ubiquitous: Mount Sinai, Columbia Presbyterian, St. Vincent's, Brookdale, Coney Island—all reported above average numbers of births (the author of the article, Martin Tolchin, noted that "New York and Brooklyn Jewish hospitals reported that their number of births was normal"). As a matter of due diligence, he asked a number of sociologists and obstetricians for their comments. One sociologist explained the matter succinctly: "The lights went out and people were left to interact with one another." Others reacted with some skepticism, eager to "see the data for the entire city:" "If it should be true, I would think it is because people may have had trouble finding their accustomed contraceptives ...."

Five years later, the demographer J. Richard Udry published a three-page scholarly article with the following terse abstract:

> A comparison of the number of births in New York City nine months after the Great Blackout of 1965 with comparable periods for the previous five years shows no increase in births associated with the blackout. [11]

The method by which he arrived at this conclusion was bracingly simple. Taking November 10th, 1965 as the date of conception, he then assumed the average gestational length to be 280 days from the last menstrual period (thus 266 or 267 days from conception). He then determined, using government records on vital statistics, that about ninety percent of the babies conceived on November 10th would have been born between June 27th and August 14th. If there were an unusually high number of conceptions on November 10th, it followed that this entire period would contain a higher number of births relative to previous years. But when he took those numbers (for the previous five years), he discovered that the given period during 1966 wasn't in any way remarkable. In fact, the percentage of births during the dates in question held steady, at around 13.9 percent, from 1961 to 1966. Blackout babies, in other words, were a myth.

It would be easy to read this story as a cautionary tale about the misuse of statistics. But for us, this episode says much more about the conditions that prevail whenever statistics are used. The point of the *New York Times* article, after all, was not merely to make the case for blackout babies, but to say something about human beings and human priorities. Without the comforts of modern technology, we turn to one another—a sad commentary on our attachment to television, but a hopeful commentary on our basic predispositions and possible return. Since this idea is couched in the language of science (Tolchin spends three full paragraphs recounting the data he was able to gather from various hospitals), everyone follows suit. The sociologist speaks of people "interacting," when it is obvious to everyone that "gettin' busy" is the real state of affairs. Another requests better data; we're talking about facts here. And facts, of course, license what might in another context be understood as a cruel, anti-Semitic slur: Jewish people, alas, are not so prone to cuddling with one another when the lights go out.

Udry, too, is telling a story, and he is clearly the hero: the scientist, armed with proper understanding of the facts, triumphs over the anecdotal impressions of the untrained. After setting forth his own numbers, he adds:

> Let us not imagine that a simple statistical analysis such as this will lay to rest the myth of blackout babies. [...] It is evidently pleasing to many people to fantasy [sic.] that when people are trapped by some immobilizing event which deprives them of their usual activities, most will turn to copulation.

People, in other words, are not just untrained, but willfully delusional. They will persist in their attachment to fantasy in the face of incontrovertible evidence. Like the sociologists, he cannot even allow himself an unadorned reference to sex. The people to which he is referring "copulate." Like farm animals.

Discussions of probability and statistics often begin by restating an old bromide: "There are lies, damn lies, and statistics." This quote, which has been variously attributed to Benjamin Disraeli, Mark Twain, Lord Courtney, and probably a dozen others besides, would seem either to dismiss the entire subject as fatuous trickery, or to encourage us to look behind every curtain for falsehood masquerading as truth. We think it would be better to acknowledge that no area of mathematics is more insistently connected to narrative. The question that has to precede "Am I being lied to?" is "What story am I being told?" And for the prospective statistician: "What story am I trying to tell?"

We doubt that any reader of this book needs to be reminded that statistics has been used in the service of many dangerous ideas (the example about Jewish hospitals above recalls considerably more frightening uses of statistics only a few decades previously). But undue emphasis on the damned lies can obscure the extraordinary power of statistical methods. If they are good at deceiving, they are likewise good at yielding insights in fields ranging from biology to linguistics. The connection between statistics and narrative may well be the most important point of intersection between mathematics and the humanities. And in this case, it may be that the former has as much to glean from the latter as the other way around.

4.1 Probability

Everything about statistics depends upon the principles of probability. The modern study of this subject originated, as so many good things do, with a bar bet. Specifically, "How do you settle an *interrupted* series of bets between equally skilled players?" Phrasing the problem in modern terms, suppose that Floyd and Sam are playing a best-of-nine series, where the first one to get five points wins the overall bet. Unfortunately, someone's phone rings and the game is broken up when Floyd is leading 3–2. It wouldn't be fair to give Floyd all the money (since Sam hasn't actually lost yet), but it also wouldn't be fair simply to split the money evenly, since Floyd does have an advantage.

The problem sounds simple, but it had been discussed for well over a century before it was brought to the attention of one of the best mathematicians of the seventeenth century, Blaise Pascal (1623–1662). His answer laid the foundations for what would become modern probability theory.

First, he pointed out that the series can't last longer than four more games, since it's a best-of-nine and they've already played five. Second, he pointed out that each game has two possible outcomes; either Floyd will win, or Sam will. How many ways can these four games be divided? Pascal identified sixteen possibilities, enumerated here. (We use the notation F for a win for Floyd, and S for a win for Sam.)

F-F-F-F	*F-F-F-S*	*F-F-S-F*	*F-F-S-S*
F-S-F-F	*F-S-F-S*	*F-S-S-F*	**F-S-S-S**
S-F-F-F	*S-F-F-S*	*S-F-S-F*	**S-F-S-S**
S-S-F-F	**S-S-F-S**	**S-S-S-F**	**S-S-S-S**

In the listing above, all of the cases where F appears at least twice give Floyd his necessary two wins, and have been listed in italics. All the cases where F does not (and where S therefore appears at least three times) are listed in bold and give Sam the three wins he needs. Simple counting shows that of the sixteen cases, Floyd wins eleven, Sam five. Therefore, Floyd should win $\frac{11}{16}$ of the stake, Sam $\frac{5}{16}$.

This solution is important, not merely because it solves the problem for card players in the age of cell phones, but because it provides a structure and a process for solving other problems like it—a structure that can be formalized numerically as well. In general, the "classical" definition of the probability of an event can be written as the ratio of the number of possible cases where that event happens to the number of possible cases in total, when all cases are known to be equally likely. For example, flipping a fair coin has probability 0.5 ($\frac{1}{2}$) of yielding up heads, since there are two equally likely cases, of which one is heads; rolling a fair die and getting a five or better has probability of $\frac{2}{6} = \frac{1}{3}$, since there are two cases out of six. The probability of drawing the Jack of spades out of a deck of cards is $\frac{1}{52}$, while the probability of drawing any Jack is $\frac{4}{52} = \frac{1}{13}$.

This principle works when applied to larger numbers. In fact, that is one of the major uses for combinatorics, as discussed in section 2.5. In that section, we showed, for example, that the number of possible five-card poker hands is $\binom{52}{5}$ (i.e. $C(52,5)$), or 2,598,960). Of those hands, exactly four are royal flushes, so the probability of being dealt a royal flush is $\frac{4}{2{,}598{,}960}$ or about 0.000001539. There are 5148 different ways to be dealt a flush, so the probability is $\frac{5148}{2{,}598{,}960}$ or about 0.001980792. Finally, we also showed that there are 3744 different full houses, yielding a probability of $\frac{3744}{2{,}598{,}960}$ or about 0.001440576. As you might expect, flushes are more common than full houses or royal flushes, and the numeric probability is therefore higher: $0.001980792 > 0.001440576 > 0.000001539$. Of course, all of these possible hands are equally likely only if the dealer isn't cheating.

The caveat "when all cases are known to be equally likely" is crucial for this formulation to work. Otherwise, you can get nonsensical arguments: "Either I am Spider-Man or I am not. There are two cases, therefore the chance that I am Spider-Man is $\frac{1}{2}$." Yet even in situations where the underlying case structure is not known or not uniform, we can still make some general observations about probability.

- All probabilities can be expressed as a number between zero and one (inclusive)—a statement that formalizes our intuitive notions of uncertainty using the properties of numbers.
- If A is more likely to happen than B, then the probability of A is greater than the probability of B.
- An experimental outcome with probability 1 will always happen. An experimental outcome with probability 0 will never happen. And in general, an event with probability $\frac{x}{y}$ will happen, in the long run, about x times out of y.
- Because probabilities are numbers, they can be manipulated using the rules of ordinary arithmetic.

This way of thinking (classical probability theory) is also called "frequentist probability." Under this framework, a statement of probability is a statement of long-term frequency—a description of what happens if you repeat an experiment over and over again. For this reason, frequentist probability is sometimes called instead the "physical" probability, since it represents the physical results of a repeated experiment. There are various kinds of interpretations we might give to the numbers involved, but the basic methods of manipulating them remains the same.

Second die		1	2	3	4	5	6
First die	1	2	3	4	5	6	7
	2	3	4	5	6	7	8
	3	4	5	6	7	8	9
	4	5	6	7	8	9	10
	5	6	7	8	9	10	11
	6	7	8	9	10	11	12

Table 4.1: Possible outcomes of throwing two dice

4.1.1 Manipulating Probabilities

The two basic rules for manipulating the numbers involved with a statement of probability can be stated fairly simply. The probability of a **disjunction** ("or") of two separate events is simply the sum of their separate probabilities. The probability of a **conjunction** ("and") of two independent events is the product of their separate probabilities.

What is the probability of rolling a 2 (snake eyes) on two dice? We can't simply say that there are 11 possibilities (2,3,4,...,12) so the probability is $\frac{1}{11}$, because the possibilities are not of equal likelihood (there are more way to roll a 7 than there are ways to roll a 2). We can observe, however, that the chance of rolling any given number on a die is always the same: one in six. The only way to roll a 2 on two dice is to roll a 1 on the first die and then to independently roll a 1 on the second die. Since the probability of rolling a 1 is $\frac{1}{6}$, the probability of rolling a 1 and then rolling another 1 is $\frac{1}{6} \cdot \frac{1}{6}$, or $\frac{1}{36}$.

This argument applies equally to any specific number; the chances of rolling double fours, or a five on the first die and a one on the second, and so forth, is equally $\frac{1}{36}$. We could even make a table of the possible outcomes (table 4.1), which would show us each of the possible and equally likely outcomes. Notice that in this table, exactly one entry shows a total of 2, so the probability of rolling a 2 is one in six ($\frac{1}{36}$) by the rules above.

We can also observe that there are exactly two ways of rolling an 11—either a five on the first die and a six on the second, or vice versa. With two possible outcomes, the probability is $\frac{2}{36}$, or $\frac{1}{18}$. But notice that we can get the same answer in another way. The probability of rolling a five and a six is $\frac{1}{36}$, while the chance of rolling a six and a five is also $\frac{1}{36}$. According to the addition rule, the chance a five and a six **or** a six and a five is therefore $\frac{1}{36} + \frac{1}{36}$, or $\frac{1}{18}$ (as before). This set of likelihoods forms the basis of the casino game craps.

But did you see the card we palmed (to mix metaphors for moment)? Tricks like this represent a common (and quite subtle) way of misrepresenting probability (and by extension, statistics). The addition and multiplication rules only work under certain conditions, and although we told you about the conditions, we didn't check to make sure they held when we did our arithmetic. In particular, adding probabilities this way only works when the relevant event sets are **disjoint**, when they can't both be true. So the events "I threw a one" and "I threw a two," or even "I threw an odd number" and "I threw an even number" are disjoint. But "I threw an odd number" and "I threw a number less than four" are not—if I throw a 3, that number is both odd and less than four.

Outcome	=1?	=2?	=1 or =2?	odd?	even?	odd or even?
1	yes	no	yes	yes	no	yes
2	no	yes	yes	no	yes	yes
3	no	no	no	yes	no	yes
4	no	no	no	no	yes	yes
5	no	no	no	yes	no	yes
6	no	no	no	no	yes	yes
Probability	$\frac{1}{6}$	$\frac{1}{6}$	$\frac{2}{6}$	$\frac{3}{6}$	$\frac{3}{6}$	$\frac{6}{6}$

Table 4.2: Addition of disjoint probabilities

So we can say that the probability of throwing a 1 **or** a two is (numerically) the probability of throwing a 1 plus the probability of throwing a 2. The probability of throwing an odd number **or** an even number is the probability of throwing an odd number plus the probability of throwing an even number. But the probability of throwing an odd number **or** throwing a number less than four is not the sum of the two relevant probabilities. We can confirm this with a truth table, as in tables 4.2 and 4.3.

The first lesson in probability, then, is to be careful—and to watch to make sure others are being careful. The second lesson is that we can modify the addition rule slightly to cover the case where the two possibilities overlap. As modified, this becomes known to mathematicians as the **inclusion/exclusion principle**:

> The probability of a disjunction of event A **or** event B is equal to the probability of event A plus the probability of event B, minus the probability of a conjunction of

Outcome	odd?	<4?	odd or <4?
1	**yes**	**yes**	yes
2	no	yes	yes
3	**yes**	**yes**	yes
4	no	no	no
5	yes	no	yes
6	no	no	no
Probability	$\frac{3}{6}$	$\frac{3}{6}$	$\frac{4}{6}$

Table 4.3: Addition of nondisjoint (overlapping) probabilities

both A and B.

Notice that this does hold for table 4.3; the probability of getting an odd number or a number less than 4 is equal to $\frac{3}{6} + \frac{3}{6}$ minus the probability ($\frac{2}{6}$) of an overlap as in outcomes 1 and 3.

This is more commonly written using the notation for sets introduced earlier. In general, the expression $P(A)$ represents the probability of A, where A is really a *set* of events. (Of course, a singleton set has only one member, so the set can be a single event). The idea of A **or** B can be expressed via set union ($\cup$) and the idea of A **and** B via set intersection ($\cap$). We can thus say that:

$$P(A \cup B) = P(A) + P(B) - P(A \cap B) \tag{4.1}$$

A similar caveat holds for the multiplication rule. For it to work, the two events need to be **independent**; they can't have any influence or connection to each other. For example, two separate rolls of two separate dice are independent. However, rolling a 1 is not independent from rolling an odd number on that same die—if I roll a 1, I have to have rolled an odd number (since one is odd), or alternatively, if I roll an even number, it can't have been a 1. In the case where events A and B are independent, we have

$$P(A \cap B) = P(A) \cdot P(B) \tag{4.2}$$

But how about for dependent events?

4.1.2 Conditional Probability

Well, the math gets a little more complicated (and for this reason, questions of independence arise all the time in the interpretation of statistics and probability). Specifically, we need now consider the idea of **conditional probability**—the idea of the probability of B "given that" something else (like A) has already happened.

So let's revisit table 4.1, but concentrate specifically on one row only. For ease of explanation, we'll let f denote the outcome of the first die, and s the outcome of the second, so the expression $P(f = 4)$ is "the probability that the first die rolled a 4."

In table 4.4, we see the fourth row highlighted, representing the case where $f = 4$. If we know that the first die *has* rolled a four, then we know that whatever the outcome is must be in that row, and that a total of two (for example) is strictly impossible. Within that row, there are only six possible outcomes, and the chance of (for example) rolling a 10 is $\frac{1}{6}$.

We can thus distinguish between "the probability of rolling a 10

s		1	2	3	4	5	6
f	1	2	3	4	5	6	7
	2	3	4	5	6	7	8
	3	4	5	6	7	8	9
	4	**5**	**6**	**7**	**8**	**9**	**10**
	5	6	7	8	9	10	11
	6	7	8	9	10	11	12

Table 4.4: Possible outcomes of throwing two dice

total" (which can be seen to be $\frac{3}{36}$) and the "the probability of rolling a 10 total, *given that the first die rolled a 4*" (which is $\frac{1}{6}$). The chance is thus substantially better if conditions are right and we are allowed to know those conditions. On the other hand, if the first die was a 1 (the first line of the table), there is no chance whatsoever of getting a 10 total. This distinction nicely illustrates the idea of conditional probability; if we have prior knowledge that restricts the possible **event space** (the set of all possible outcomes), we can "condition" our probability calculations based on that knowledge.

The notation used for this—which is entirely ubiquitous in discussions of probability across disciplines—is $P(B|A)$, read "the probability of B given A." For example, $P(f+s=10)$ (the probability that the sum of the two rolls is 10) is different from $P(f+s=10|f=4)$ (the probability that the sum of the two rolls is 10, given that the first die roll was a 4), and different from $P(f+s=10|f=1)$, as we have just seen. Armed with this notation, we are prepared to give a more detailed analysis of the idea of independence.

Two events A and B are **independent** if and only if the fact that A happened has *no effect* on the probability of B and vice versa; in other words $P(A) = P(A|B)$ and $P(B) = P(B|A)$. In general, the probability that both A **and** B happen is given by

$$P(A \cap B) = P(A) \cdot P(B|A) = P(B) \cdot P(A|B) \tag{4.3}$$

and, of course, if A and B are independent, this reduces to

$$P(A \cap B) = P(A) \cdot P(B) = P(B) \cdot P(A) \tag{4.4}$$

as previously illustrated.

Some other useful facts about conditional probabilities:

- The technical formula for conditional probability is $P(A|B) = \frac{P(A \cap B)}{P(B)}$.
- There is no necessary relationship between $P(A|B)$ and $P(A)$, since event B can make A either more or less likely. Therefore, $P(A|B)$ can be greater than, less than, or equal to $P(A)$.
- $P(A|B)$ and $P(A|\bar{B})$ (the macron is used to indicate that B does not hold, see section 2.1.2) describe disjoint events, so they can be summed directly. This gives rise to the so-called the **law of total probability**, which can be expressed as:

  $$P(A) = P(A|B)P(B) + P(A|\bar{B})P(\bar{B}) \tag{4.5}$$

  In essence, this rather tedious formula states that the chances of A being true are equal to the chances of A being true and B

being true, plus the chances of A being true and B not being true. Since B can't be true and false at the same time, we just add the probabilities in accordance with the rules we've already seen.

Armed with these rules (and with what we've said in previous chapters), it should be apparent that one can develop an algebra of probabilities and work with them using the ordinary arithmetic operations as defined on numbers. Such a framework in turn allows us to prove what intuition already suggests: that if, for example, P has a higher probability than Q, then P will not occur any less frequently than Q.

4.1.3 Bayesian and Frequentist Probabilities

Bayes's Theorem

Unfortunately, the probability framework developed above doesn't capture many of our philosophical notions of probability. For example, what was the "probability" of the Brazilian national team winning the World Cup in 2006? At the start of the competition, they were heavily favored (a judgment borne partially by statistics, and partially by a general, intuitive sense of which team was better). We know in hindsight that they didn't win, but since the 2006 World Cup is a one-off event, there's no way to make statements like "they will win x out of y times." The frequentist or classical idea suggests what will happen over a repeating series of 2006 World Cups—which is obviously nonsensical.

But there's another way to interpret the idea of probability: not in terms of frequency, but in terms of "degree of belief." If I say that something has a probability of 0.90 (or "is 90% likely to happen"), I'm saying (among other things) that I think it's substantially more likely to happen than not to happen, even if we're talking about a one-off event. In the case of a past event, I'm saying that I wasn't very surprised that it happened, or alternatively that I was really surprised that it didn't. Of course, I wouldn't be as surprised if I said it was only 60% likely. (And specifically, if I say something like "90% likely," then I'm implicitly stating that I think a bet at 9:1 odds would be a fair bet on that event.)

This framework, where we assess probabilities in terms of expectation and belief, is sometimes called **Bayesian probability** (as opposed to frequentist or classical probability theory), after Thomas Bayes (1702–1761). To understand what this means, we have to go back to algebra.

In the previous section, we developed the notion of conditional probability—the idea of adjusting our ideas of the likelihood of an effect based on the presence or absence of a causal factor. We expressed that idea mathematically (in equation 4.3) as

$$P(A \cap B) = P(A) \cdot P(B|A) = P(B) \cdot P(A|B)$$

This, of course, is an equation. Considering it in purely algebraic terms (and using nothing more than secondary-school algebra), we can re-express it as

$$P(A) \cdot P(B|A) = P(B) \cdot P(A|B) \tag{4.6}$$

$$P(B|A) = \frac{P(A|B) \cdot P(B)}{P(A)} \tag{4.7}$$

But in doing this, we (or, rather, Thomas Bayes) have done something quite astonishing. This new theorem (called **Bayes theorem**) allows us to trace how learning new information (A) *affects our assessment of old information (B).* The value $P(B)$, often called the **prior probability,** is our assessment of the likelihood that B is true before we get a new piece of evidence. Once we learn that A is also true, we can update our prior probability to get a **posterior probability** ($P(B|A)$), the probability that B is true in light of that new fact. In other words, we are now interpreting probability, not in terms of repeated physical events, but in terms of mental events and "evidential likelihood."

Notice, also, that we could expand the formula to include the particular insights of the law of total probability, thus arriving at:

$$P(B|A) = \frac{P(A|B)P(B)}{P(A|B)P(B) + P(A|\bar{B})P(\bar{B})} \tag{4.8}$$

An unwieldy formula, to be sure, but one that captures most of what we now know about conditional probability. In practice, we simply ensure that $P(A)$ has been calculated appropriately, and proceed with the shorter version.

A Bayesian Narrative

Let's suppose that investigators are looking at a fire in a restaurant, and are trying to decide if it was arson. Specifically, they are trying to determine the probability that the fire was deliberately, as opposed to accidentally, set. From a frequentist standpoint, this question doesn't even make sense. The fire is a one-off, and it was either arson or it wasn't. But we can talk about our degree of confidence that it was arson, or alternatively about how surprised we would be to find that it was an accident. If the intrepid inspector says, "I'll give you nineteen to one that it was arson," he's implicitly stating that he is extremely (95%) confident in his conclusion.

But how does he get to that conclusion? Let's say that the investigators have learned the following facts (either from a site inspection or just from experience).

1. About 10% of the fires in this town are arson (i.e. the probability that a fire is arson is about 0.10 overall). The inspector knows this, the prior probability, even before he arrives at the crime scene.

2. There were strong signs that this particular fire burned very hot—hotter than one would expect for a typical restaurant fire, but consistent with the use of something like thermite to start a fire quickly. In fact, only about 8% of the fires in this town burn that hot. This new information will help him determine a posterior probability.

3. About 50% of the arsons in this town use accelerants like ther-

mite. In other words, the probability any given fire burns that hot *if* it were arson is about 0.50.

What can we do with all of this?

Let B be the idea that the fire was started deliberately. From fact #1, we know that $P(B) = 0.10$. Let A be the idea that the fire was very hot. This is a rare fire (only 8% of fires are that hot), so $P(A) = 0.08$. From fact #3, we know that the probability of the fire burning that hot given that it was arson is about 50%, so $P(A|B) = 0.50$. However, we, and the investigators, are interested in the probability that a very hot fire was caused by arson, or $P(B|A)$. By Bayes's theorem, we know that

$$P(B|A) = \frac{P(B) \cdot P(A|B)}{P(A)} \quad (4.9)$$

$$= \frac{0.10 \cdot 0.50}{0.08} \quad (4.10)$$

$$= \frac{0.05}{0.08} \quad (4.11)$$

$$= \frac{5}{8} \quad (4.12)$$

$$= 0.625 \quad (4.13)$$

In other words, the chances are a little over 60/40 that this particular fire was deliberately set, and the investigator should definitely do some more digging. He shouldn't offer 20:1 odds, but can probably bet 8:5 with confidence.

He has therefore updated his prior estimate of the likelihood of arson (0.10) to a higher posterior estimate (0.625), reflecting the compelling evidence from the heat of the fire. As more evidence comes in, the posterior becomes the new prior, and he can update this new prior even further, until all the evidence has been accounted for.

This simple analysis, as you might guess, has important sociological consequences. Suppose an eyewitness comes forward, claiming to have seen a suspicious-looking person fleeing the scene, and describes the alleged perpetrator as a member of a particular minority group. How much should one trust the eyewitness?

Eyewitnesses are known to make errors, and we might want to test the witness. If after the tests, we confirm that she is not that reliable, the following facts are known:

1. About 5% of the criminals of this town are of that minority.

2. The witness is about 70% accurate. That is, about 70% of the time, she can correctly identify the race of a member of the minority group, and equally importantly...

3. about 70% of the time, she can correctly identify the race of someone who is not of that group (or alternatively, about 30% of the time, she mis-identifies someone as that minority).

Letting *A* be "the criminal is a member of that minority group" and *B* be "the witness identified the criminal as that minority," we have the following:[1]

[1]Remember from section 2.1.2 that $\bar{A}$ means "everything that isn't *A*." In this

1.
$$P(A) = 0.05 \tag{4.14}$$

2.
$$P(B|A) = 0.70 \tag{4.15}$$

3.
$$P(B|\bar{A}) = 0.30 \tag{4.16}$$

Using, again, the law of total probability, we can calculate the probability $P(B)$ as

$$\begin{aligned} P(B) &= P(B|A)P(A) + P(B|\bar{A})P(\bar{A}) && (4.17)\\ &= 0.70 \cdot 0.05 + 0.30 \cdot 0.95 && (4.18)\\ &= 0.32 && (4.19) \end{aligned}$$

In other words, this witness would probably identify about 32% of the population as belonging to this minority, mostly in error. You can probably see where we're going with this.

What's the chance, then, that the criminal is actually a minority, taking into account the witness's statement? Well,

case, it means that A isn't true: the suspect isn't a member of that minority group.

$$P(A|B) = \frac{P(B|A)P(A)}{P(B)} \tag{4.20}$$

$$= \frac{0.70 \cdot 0.5}{0.32} \tag{4.21}$$

$$= .1093 \tag{4.22}$$

The investigators can rightly conclude that the witness is helpful, but not overwhelmingly so. Given the relative rarity of this minority, it is eight times as likely that the witness made a mistake as got it right. The witness's statement is evidence, but hardly conclusive.

As we mentioned, one advantage of the Bayesian framework, is that it can be applied repeatedly to take advantage of other, additional evidence as it comes in. The posterior probability becomes the prior probability in the next round; we've updated our estimate from about 1-in-20 to about 1-in-9. We can now continue to update as the reports trickle in. For example, a study of the scene of the crime found some bloodstains of a rather unusual type. In particular, this type is rare in the population at large, but relatively common in the minority. (Genetically, this isn't that uncommon—Peruvian Indians, for example, have near 100% type O blood, so if a bloodstain in Lima tests as type B, it's probably not from a Peruvian Indian.)

As before, A is the idea that the criminal is of this minority, and our current estimate (including the witness' statement) is that $P(A) = 0.11$. Let C be the chance that the criminal has this type of blood.

1. About 5% of the majority have this type: $P(C|\bar{A}) = 0.05$

2. About 50% of the minority have this type: $P(C|A) = 0.50$

3. Therefore, by the law of total probability,

$$P(C) = P(C|A)P(A) + P(C|\bar{A})P(\bar{A}) \tag{4.23}$$
$$= 0.50 \cdot 0.11 + 0.05 \cdot 0.89 \tag{4.24}$$
$$= 0.10 \tag{4.25}$$

4. Using Bayes's theorem, we find that

$$P(A|C) = \frac{P(C|A)P(A)}{P(C)} \tag{4.26}$$
$$= \frac{0.50 \cdot 0.11}{0.10} \tag{4.27}$$
$$= 0.55 \tag{4.28}$$

This new information makes the chances better than even than the arsonist is a minority. That's probably enough to let the investigators change how they do the investigation. In this case, probability is being used not to prove that the case was an instance of arson, but to guide the investigation itself more fruitfully.

Another important example of such statistical guidance occurs in disease testing—especially when it comes to rare diseases. In general, no test is perfect, and will always have some false positives and some false negatives. For example, a typical (i.e. cheap) test for hepatitis might be 95% accurate. Now consider the issue of a person whose test has just come back positive. What's the chance that he actually has hepatitis?

As we can see from the discussion above, it's not just 95%; it depends on the actual prevalence of hepatitis in the population. Again, we approach this from a Bayesian perspective. The reasoning should be familiar at this point.

1. Assume that 0.1% of the population has hepatitis : $P(A) = 0.001$

2. 95% of the positive tests are accurate : $P(B|A) = 0.95$

3. 5% of the negative tests are inaccurate : $P(B|\bar{A}) = 0.05$

4. By the law of total probability,

$$\begin{aligned} P(B) &= P(B|A)P(A) + P(B|\bar{A})P(\bar{A}) && (4.29) \\ &= 0.95 \cdot 0.001 + 0.05 \cdot 0.999 && (4.30) \\ &= 0.0507 && (4.31) \end{aligned}$$

   Therefore, the chance of any particular test coming back positive is a little over 5%, reflecting the chances of a false positive.

5. Using Bayes's theorem, we find that

$$\begin{aligned} P(A|B) &= \frac{P(B|A)P(A)}{P(B)} && (4.32) \\ &= \frac{0.95 \cdot 0.001}{0.0507} && (4.33) \\ &= 0.0187 && (4.34) \end{aligned}$$

Thus, we conclude that the person who gets a positive test result has a little less than 2% chance of having hepatitis. This has raised his chances by something like 19-fold, and definitely justifies more extensive testing and observation, but the odds are still 50:1 that this is a false positive, simply because of the rarity of the original condition. As you can imagine, this is a matter of grave serious concern for those setting health policy involving the use of wide-scale screening.

Bayes's theorem can be used to formulate numeric measures for belief and confidence, and help guide the interpretation of new evidence. It can provide firm numeric guidance for why we shouldn't take eyewitnesses or positive blood tests too seriously, or to keep us from underestimating the significance of a rare finding. It cannot, however, make decisions. For informed decision making, we need a little more scaffolding.

4.2 Probability and Statistics

If probability is a mathematician's way of quantifying what hasn't happened yet, statistics is the way mathematicians quantify what has already happened. The subfield known as **descriptive statistics** reveals this function in its most basic form, since it is mostly concerned with summarizing data. The simple graphs shown in figure 4.1 illustrate this role. These graphs summarize in a few lines the income distribution for literally hundreds of thousands of (imaginary) people. Rather than poring through hundreds of pages of census data, we can turn to the graphs, which summarize the demo-

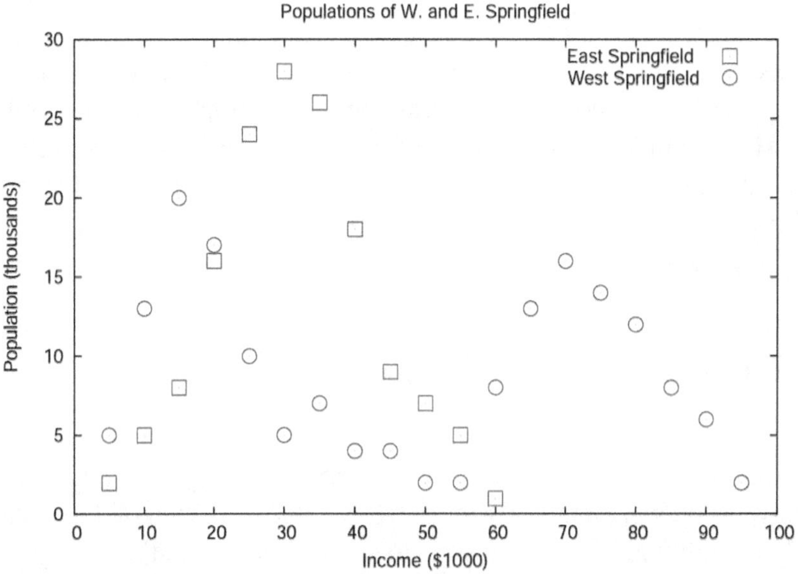

Figure 4.1: Income distribution in two imaginary towns

graphics such that interested people can see the large-scale pattern very quickly. For example, if I wanted to open a store focusing on luxury goods for the $40–60,000 annual income market, where do I open it? (Personally, we'd put it in East Springfield, since West Springfield doesn't have a lot of people in that income group.)

Our graph represents only one possibility for displaying data; depending upon the kind of data you're working with, you might decide to use **pie charts, histograms, scatter diagrams, Pareto charts, frequency polygons**, or **stem-and-leaf plots**—all of which can be generated using standard plotting tools or from within statistical software packages. The visualization of quantitative data is, in fact, an essential element of the study of statistics, with a vast literature unto itself. We do not go deeply into the subject here (sticking mostly to simple graphs), but careful study of the conventions of statistical graphics is obligatory for anyone who intends to tell stories with data. Fortunately, the subject itself is riveting—and occasionally beautiful.

4.2.1 Normal Distribution

Figure 4.2 shows a histogram of the chest sizes of 5738 Scottish militiamen, gathered in the early nineteenth century.[2] Perhaps obviously, if you were a military tailor in Scotland, this kind of data could help inform you as to which sizes to stock. Looking at the histogram, for example, we see that only a few were scrawny enough to fit into a 33-inch uniform (you will only sell a few of those). The same holds for the barrel-chested 48-inch uniforms. Your best-selling size will be the medium-sized 39- or 40-inch.

Notice that the distribution in figure 4.2 is mostly symmetrical, with a large mass in the middle of the graph, a vague bell-shaped flaring, and two tails that drift off into irrelevance. Lots of things—especially biological and social things with complex causes—distribute in this way. For this reason, the distribution has come be called the

[2]These, and other crucial data points, can be found at the (delightfully named) Data and Story Library at CMU (`http://lib.stat.cmu.edu/DASL/`), from which most of the rest of the data in this section were taken.

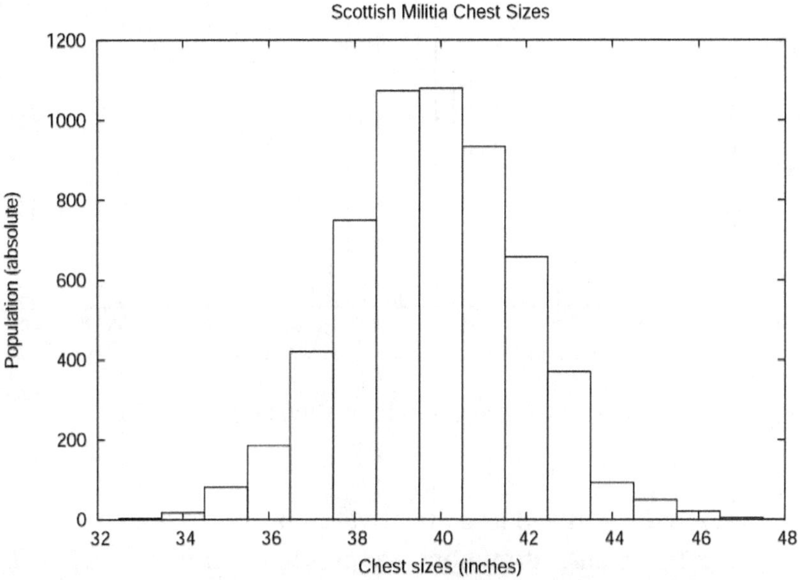

Figure 4.2: Chest sizes of Scottish militiamen

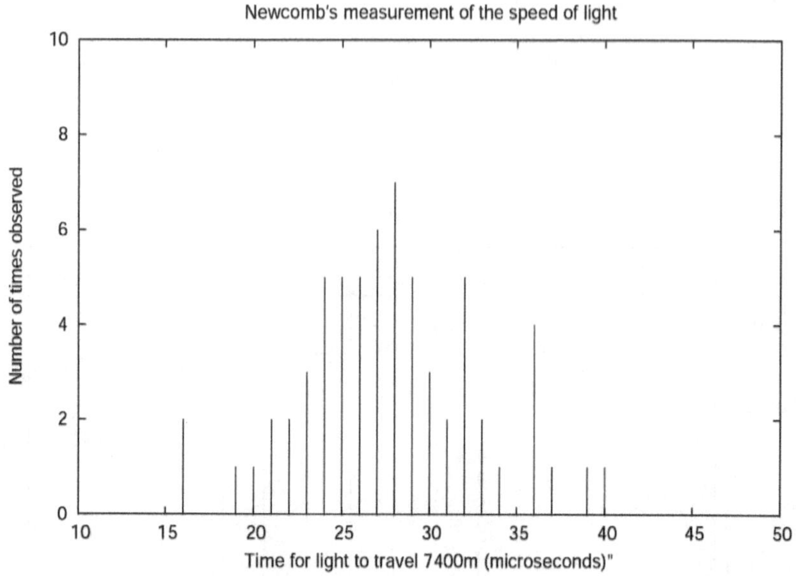

Figure 4.3: Measurements of the speed of light

normal distribution and its distinctive shape, the **bell curve.** Other examples of normal distributions are presented in figures 4.3–4.7.[3] Technically, these are only approximations of normal distribution. A true normal distribution—much like a perfect triangle—is a mathematical idealization.

[3]The temperature data is taken from `http://www.amstat.org/publications/jse/v4n2/datasets.shoemaker.html`, which also offers an interesting explanation for some of the apparent patterns in this dataset.

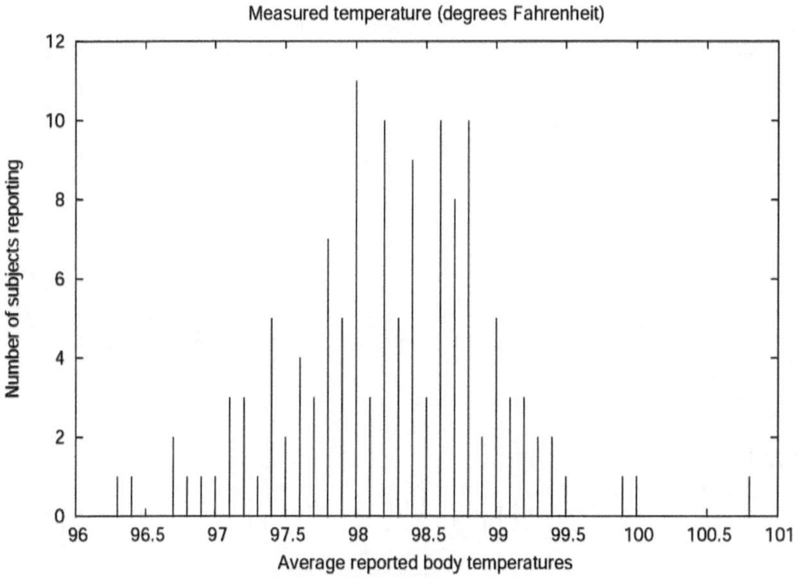

Figure 4.4: Measurements of body temperature

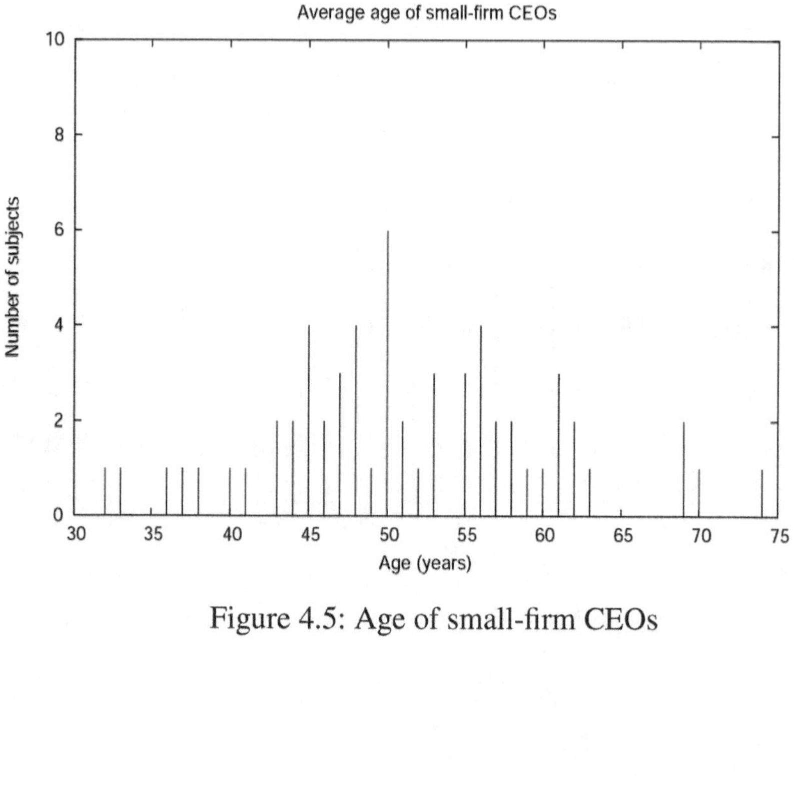

Figure 4.5: Age of small-firm CEOs

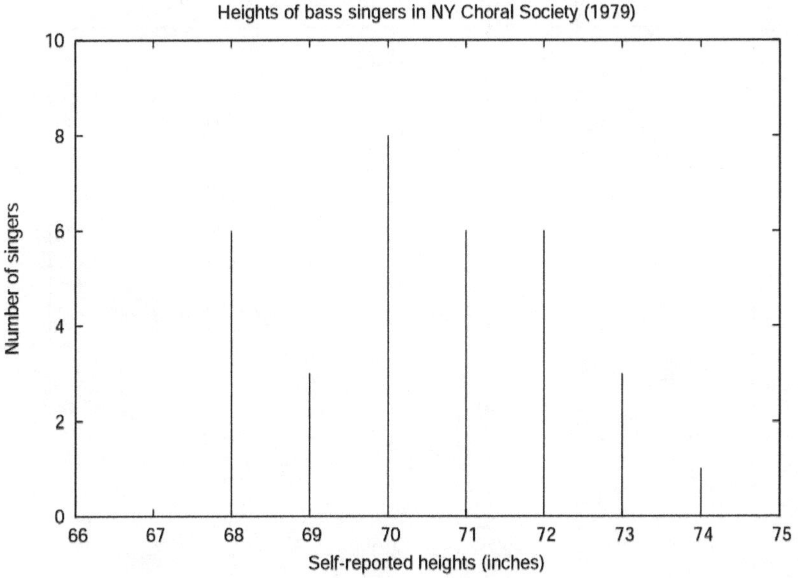

Figure 4.6: Heights of bass choristers

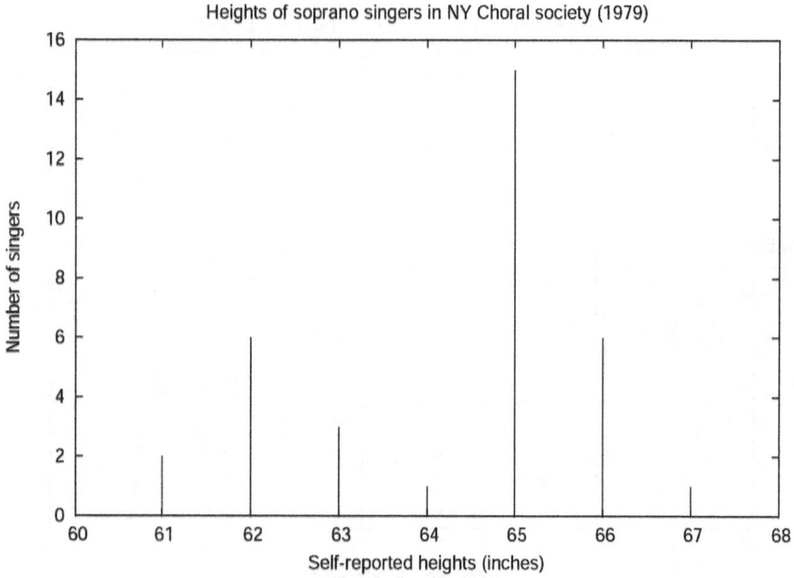

Figure 4.7: Heights of soprano choristers

There are both theoretical and practical reasons why the normal distribution is important. As we said, normal distributions are very common. But there is a good theoretical reason for this. The **Central Limit Theorem**, one of the most important theorems in statistics, says that if you take enough data from any distribution and calculate the sum of the data (or use the sum in some way, as when taking the average), the resulting sums will themselves be distributed normally.[4]

Of course, not all data is normally distributed. In Figure **??**, there are more people to the right of the average than to the left, reflecting an **asymmetry** in the distribution. The West Springfield income data is another, albeit fictitious example. Income data distribution is, in fact, rarely normal (there are more people making $100,000 more than average than there are making $100,000 less than average, for example), but the West Springfield data goes all the way to downright bizarre. And one important thing that a sociologist would want to know about that income data is why the incomes are distributed so unevenly. Perhaps there is a huge degree of racial and social inequality in that particular community? The absence of anything even approximating the normal distribution is a key indicator that something unusual is happening.

[4]It would be a grave mistake, of course, to interpret the Central Limit Theorem—or the ubiquity of the normal distribution—as some kind of irrefragable law of the universe that expresses how things should be in a deterministic sense. The "moralization" of the bell curve in this way has been a part of some of the most appalling social theories in the history of the world. Damn lies are one thing; dangerous pseudo-scientific nonsense is another.

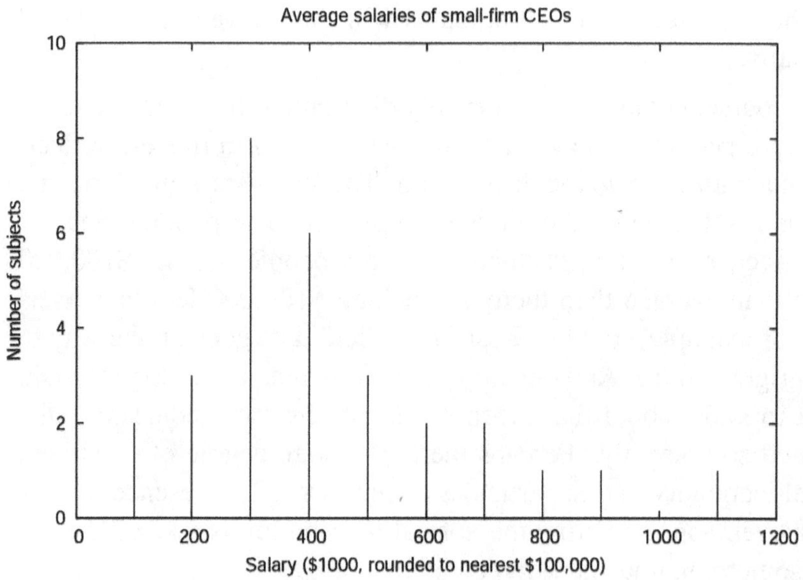

Figure 4.8: Salaries of small-firm CEOs

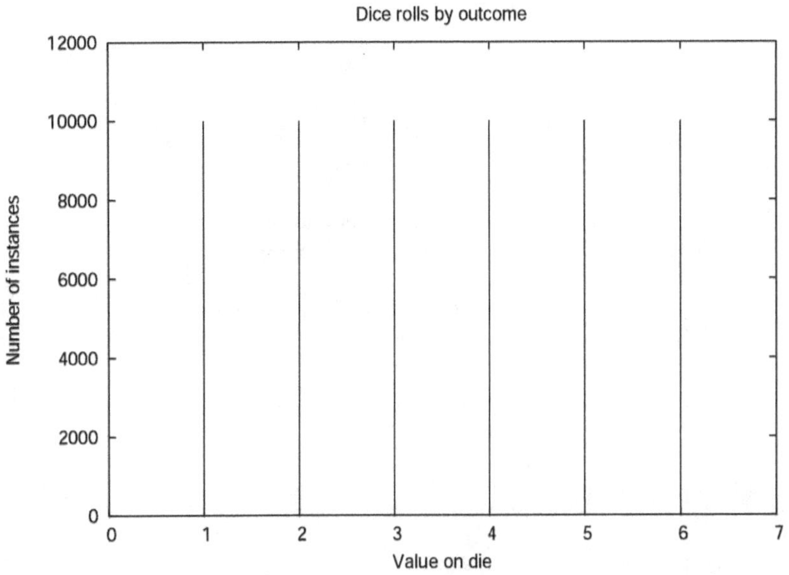

Figure 4.9: 60,000 rolls of an unrealistically fair die

4.2.2 Other Distributions

There are a few non-normal distributions that come up often enough that they deserve special names and mentions. One simple example of such a distribution is the **uniform distribution**, where each item is equally likely, as in figure 4.9. This kind of distribution arises in situations like (fair) die rolls or roulette wheels spins; indeed, if the roulette wheel at the local casino does not implement a uniform distribution, then someone's probably cheating.

Another important distribution with its own name is the **Pois-**

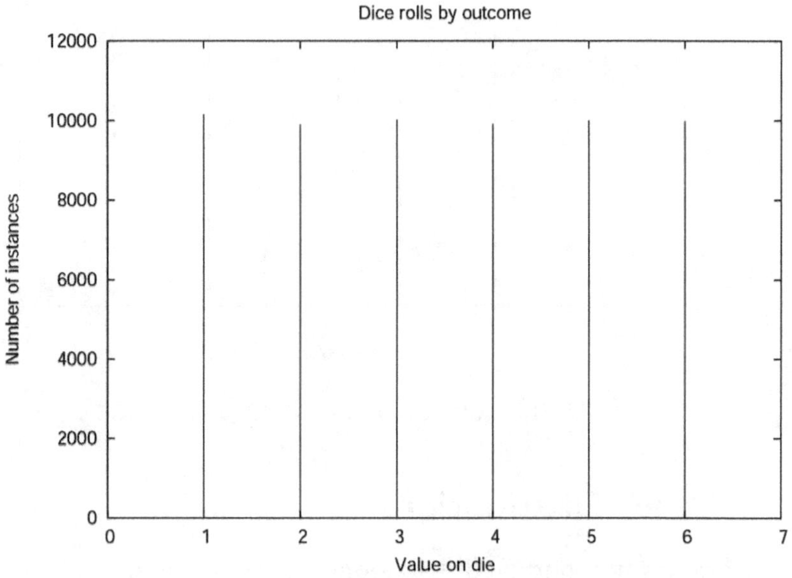

Figure 4.10: 60,000 simulated rolls of a realistically fair die

son distribution, named after Siméon-Denis Poisson (1781–1840). The classic formulation is that of Ladislaus von Bortkiewicz (1868–1931), who identified this distribution in the statistics of army officers kicked to death by horses. Obviously, there can never be fewer than zero deaths in such a manner, and the statistics suggest an average (based on von Bortkiewicz's data, see figure 4.11) of fewer than one such death per army corps per year. On the other hand, it is possible (although highly unlikely) that there could be ten, twenty, or a hundred such deaths in a year. The Poisson distribution, unlike the binomial or normal distribution, is thus very asymmetric, as can be seen in figure 4.12 as well[5].

The Poisson distribution is used as a model of rare events such as lighting strikes, deaths in childbirth, or earthquakes. It applies when we have rare and independent events for which the (low) probability is known, and we wish to count how many events are likely to occur in a given period of time. For example, the probability of having a baby at any particular time is small (in this data set, there are fewer than two babies per hour on average), but some clustering by chance is likely to happen. This is an important model for understanding questions like "How many cashiers does a store need to hire to keep customer wait time at a reasonable level" (a surprisingly difficult question that is dealt with in the rarefied mathematical field known as "queuing theory") or to the recognition of rare patterns (such as appears with fraud detection in credit card usage).

Finally, the Springfield data are another example of a non-normal,

[5]Data taken from `http://www.amstat.org/publications/jse/secure/v7n3/datasets.dunn.cfm`

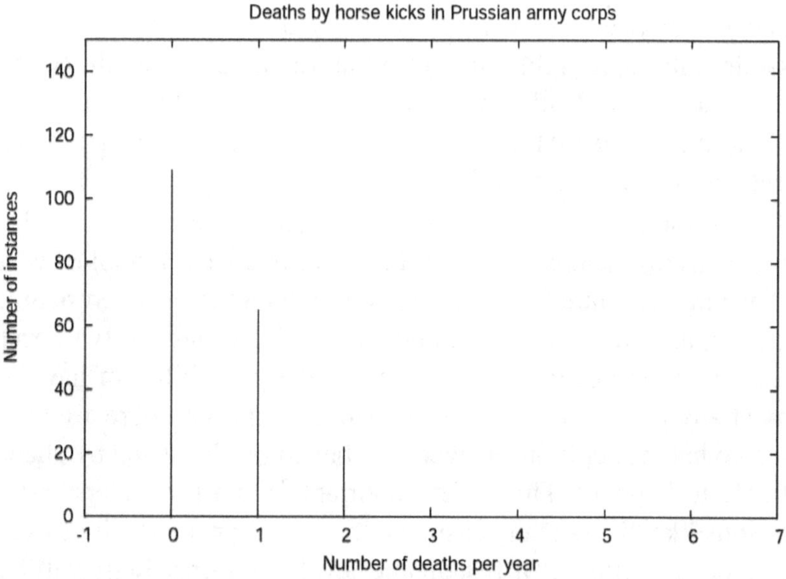

Figure 4.11: Number of soldiers kicked to death by horses, per year.

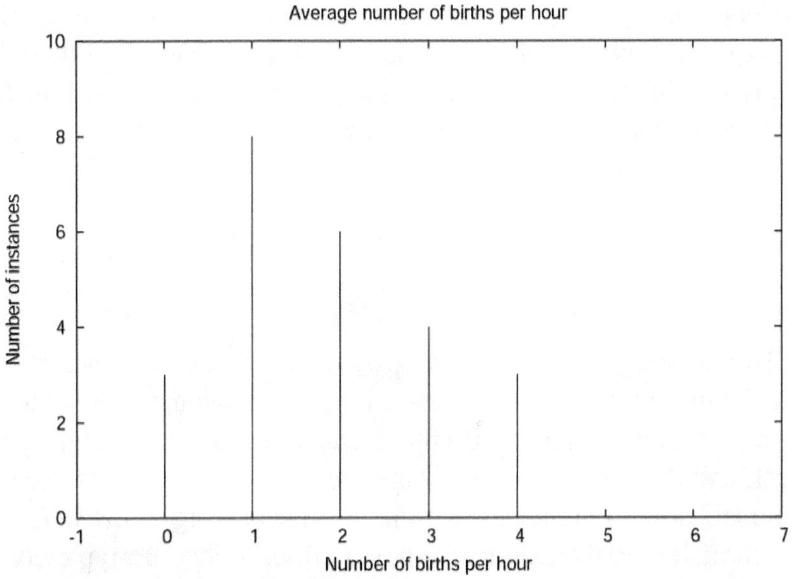

Figure 4.12: Number of births per hour as measured over a day

non-uniform distribution. This type of distribution (figure 4.13) is usually called a **bimodal distribution**. "Mode," as we will see in the next section, is simply the technical term for the most common element in a dataset. It often occurs at the peak of a distribution like the normal or Poisson distribution. In this data set, there are two such peaks. This kind of distribution is very unusual and often occurs when you are conflating data of two separate kinds. (For this example, we rolled 30,000 "fair" pairs of dice and 30,000 unfair triples—which should indicate our devotion to the subject). The first peak is the most common value for the fair dice (7), the second for the weighted dice.

Human height data, for example, will often be bimodal, reflecting the different averages of men's heights and women's heights in a group of mixed gender, with the average man's height creating one peak and the average woman's creating the other. (You can see for yourself how to create such a mixture by looking at the height of all choristers in the examples above). Book prices are typically bimodal, representing the difference in price between paperbacks and hardcover (or between new and used, as in a college bookstore). In fact, you might even see a "tetramodal" distribution of prices, with new hardcovers, used hardcovers, new paperbacks, and used paperbacks each contributing their own price peak—thus showing the utter futility of making sweeping generalizations about textbook marketing in general.

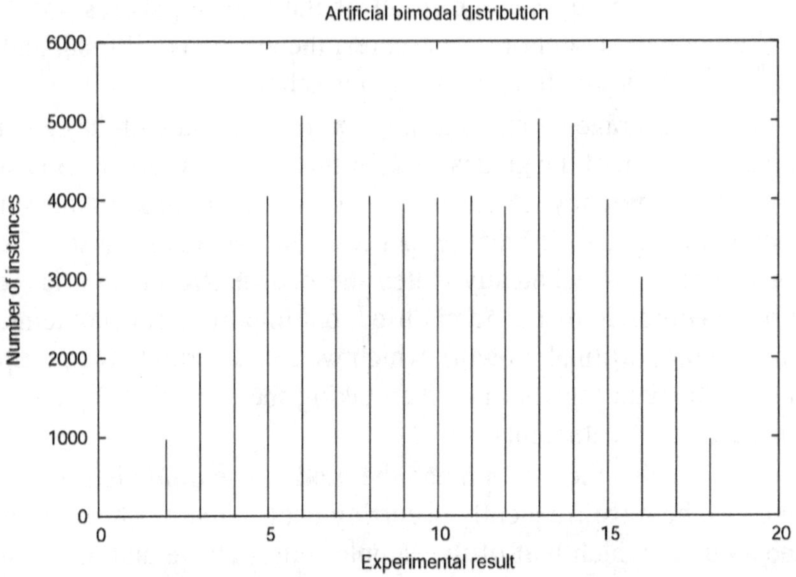

Figure 4.13: Example of a bimodal distribution

4.2.3 Central Tendency

Since distribution can take on such weird shapes, it's often useful and informative to have numeric descriptions to help in comparing different groups. At least one of these descriptions you're already familiar with: the good, old-fashioned **average**. We calculate the "grade average" by adding up all of the relevant scores and then dividing by the count of those scores; the average of 70, 80, and 90 is $\frac{70+80+90}{3}$, as we all learned in grade school.

But an "average" is really an approximate measure of the center of the distribution of the grades, and for this reason is called a **measure of central tendency**. There are actually a number of such measures, each with slightly different properties. The average we just defined, for example, is technically called the **arithmetic mean**. There is also a **geometric mean** (rarely used, but important for problems involving proportional growth), which we can determine by multiplying all the data together and then taking the N-th root (where N is the number of datapoints).

There is also the **median** and the **mode**. The **mode** is simply defined as the most frequently occurring data element. The **median** is the value at which half of the sample scores above and half below. For an example of how these can differ, consider the case of a limousine containing a wealthy industrialist making $1,000,000/year, her private secretary making $40,000/year, a driver making $30,000 and two security guards making $20,000. The mean (average) income of the people in the car is $\$\frac{1,110,000}{5}$, or $370,000, but that hardly seems an accurate way to describe the central tendency.

The modal income is simply the most common salary, the $20,000 that the security guards make. If you picked one of these people at random and asked how much they make, that would be the most common—the "average"—answer. The median income is the driver's $30,000, since two people make more than that and two make less (the driver thus represents the "midrange" of salaries). These three numbers (mean, mode, and median) are all arguably averages, but they are all different values, and mean different things.

All of this demonstrates that the distribution of incomes in the car is not "normal" in the sense discussed above. The millionaire industrialist is an **outlier**; her income is so far out of the range of the rest that she distorts the overall mean. With normal distributions, all three averages are the same, but the further things get from normal, the more they differ. Formally, we can say that the income distribution is **skewed** and asymmetric, and precisely because of this skewing, the different "averages" mean different things. As you might guess, choosing the measure of central tendency most likely to get your point across (as opposed to the measure that most accurately describes the central tendency) represents one of the more damnable ways to lie with statistics.

What is the "average" income in your home town? If I want to persuade you of how wealthy my town is (perhaps I'm selling real estate), I might want to overstate it. At the same time, I might want to understate that wealth when I'm trying to persuade the government that new taxes will adversely affect our poverty-ridden community. But is the income "normally distributed?" Suppose that in my mostly middle-class town, there's one billionaire. This is a pretty good description of the street in Oxfordshire where one of the authors once lived. The "billionaire" was Richard Branson (of

Virgin Records fame). The median income was probably less than £30,000, but the "mean" income—because of the influence of the outlier—might have been ten times that. Living near Richard Branson will inflate the mean income of the area, but not the median.

This can be seen more formally in the data in table 4.5. The left side of the table shows the data for the incomes of thirteen people in a pretty good approximation of a normal distribution; the modal (most common) value ($30,000) is in the middle of the group, and the incomes are distributed symmetrically around it. The median income happens at the exact center of the list (with Ginger) at $30,000, and the total sum of the incomes is $390,000, yielding a mean income of $30,000. The right side, however, shows the effect of our outlier. Richard has had a thirteen-million dollar windfall, which pushes the total income up to $13,270,000 and the mean to just over a million dollars, despite the fact that only one person in the entire set makes that much. Can we really have a situation where more than 90% of the population is below "average?"

4.2.4 Dispersion

The real lesson, of course, is that "average" by itself isn't all that meaningful. The notion of skewness has already been introduced as a measure of how asymmetric a distribution is, and (implicitly) as a measure of how meaningful the "average" is as a single concept. But even in a perfectly symmetric distribution, a well-defined concept such as the mean still loses information. (One could argue that this is implicit in the idea of summary statistics; a summary always loses information. But how does one know what information is being lost,

Person	Income
Angela	$10,000
Bertram	$20,000
Catherine	$20,000
David	$20,000
Elizabeth	$30,000
Frank	$30,000
Ginger	*$30,000*
Harry	$30,000
Inez	$30,000
Jeremy	$40,000
Katherine	$40,000
Lawrence	$40,000
Richard	$50,000
mode	$30,000
median	$30,000
mean	$30,000

(a) Approximately normal distribution

Person	Income
Angela	$10,000
Bertram	$20,000
Catherine	$20,000
David	$20,000
Elizabeth	$30,000
Frank	$30,000
Ginger	*$30,000*
Harry	$30,000
Inez	$30,000
Jeremy	$40,000
Katherine	$40,000
Lawrence	$40,000
Richard	$13,050,000
mode	$30,000
median	$30,000
mean	$1,030,000

(b) Richard is an outlier

Table 4.5: How an outlier can change the mean without shifting the median

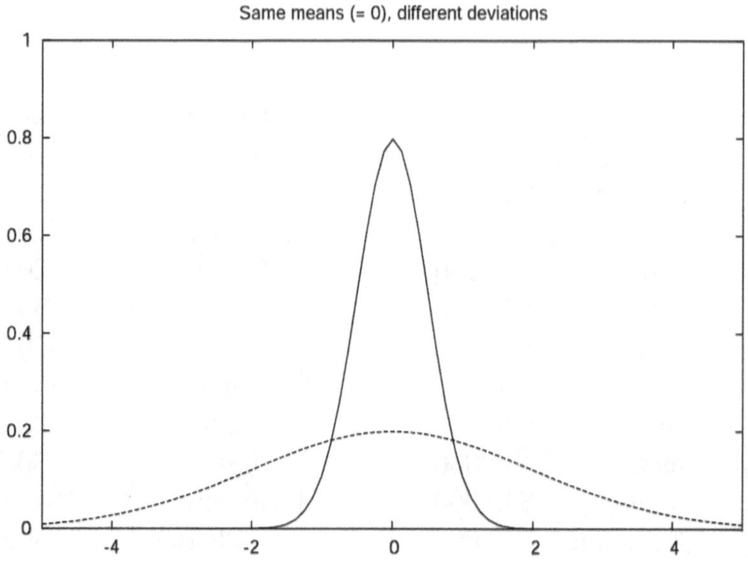

Figure 4.14: Two different bell curves

and how meaningful that loss is?)

For example, figure 4.14 shows two different bell curve distributions. The total area of both curves is the same, and similarly both curves are symmetric around zero. Although the means (and medians and modes) are thus the same in both cases, the curves are quite different. In particular, the shorter curve is also fatter—a larger percentage of the total data lies between imaginary vertical lines at +1 and -1. The **dispersion** around the mean is greater.

Degree of dispersion is usually thought of in terms of **variance** and/or **standard deviation**. We calculate both by computing the mean distance *from the overall mean of the dataset.* Specifically, consider any data point x and consider the expression $x - \bar{x}$ (where $\bar{x}$ in this case represents the average of all the data we have). Obviously, if x is close to average, then $x - \bar{x}$ is close to zero, and if x is strongly different from the average, then x will be "distant from" zero. So if we calculate the "average" distance over all values, that will give us a measure of dispersion.

Unfortunately, this "distance" (more formally, deviation) can be either positive or negative, and if $\bar{x}$ is truly average, then it will be negative about as much as it will be positive. So in order to make sure that the "distance" is always positive, we first square the deviation. So the actual expression for the **variance** (usually written as σ^2) is the mean of the squared deviations:[6]

$$\sigma^2 = \frac{(x_1 - \bar{x})^2 + (x_2 - \bar{x})^2 + (x_3 - \bar{x})^2 + (x_4 - \bar{x})^2 + \ldots + (x_N - \bar{x})^2}{N} \tag{4.35}$$

for N data points.

In physical terms, the larger the variance, the wider the base of the bell, as can be seen in figure 4.14.

For normal distributions, there is a well-known relationship between the standard deviation, the mean, and the probability of a certain degree of difference from the mean. For example, IQ tests are

[6] The **standard deviation** σ is just the square root of the variance ($\sqrt{\sigma^2}$); this is often summarized as the expression "root mean square" or RMS. Sometimes you'll see the mean written using the Greek letter μ (m, for "mean"). Notation varies somewhat, though these symbols are fairly standard.

designed to have a normal distribution with a mean of 100 and a standard deviation of 15. In general, about 69% of the population will be within one standard deviation of the mean (if the distribution is normal), so we would expect about 15% of the population to have IQs above 115, 15% below 85, and 69% to be in the range of 85–115. This relationship only holds, however, when the distribution is normal. The mean number of legs per person, for example, is just under 2 (reflecting the fact that some, due to injury or other reason, have one leg). Since the vast majority of the population has two legs, this means that the vast majority of the population has more legs than average. But many more people have 1 leg than 3....

Our IQ test also allows us to give **percentile** ranks: if your IQ is 115, you are at the 85th percentile of the population. To speak of percentiles is, of course, to indicate that you have broken up the data into percentages. It is also possible to speak of **quartiles** and (occasionally) **deciles**. In each case, we are dividing the **range** of values (the highest value in the set minus the lowest value) into discrete groups (four groups in the case of quartiles, ten groups in the case of deciles). Such divisions are often used precisely to exclude outliers. For example, in describing central tendency in a curve with extreme high or low values, it might make more sense to compute the measures of central tendency for the **interquartile range** (the two "inner" groups of values within a set of data that have been divided into four equal groups).

4.2.5 Association

A final important concept is that of **association**, which measures the degree to which two different measures vary together. For example, taller people are heavier, larger houses cost more money, and larger cars get worse gas mileage. The most common technical measurement of association is **Pearson's product-moment correlation**—often referred to more simply as the **correlation** or **Pearson's** *r*—which measures the degree to which, if you plotted two data sets against each other, you would get a straight line. Look at figure 4.15 and notice the near-perfect line on the top, the mediocre line in the middle, and the total noise on the bottom. Calculating the correlation of all three sets would give the highest correlation in the figure on the top and the lowest on the bottom.

Two data sets that are entirely uncorrelated will have a correlation of 0.0. There are two ways to get perfect correlation. An *r*-value of 1.0 means that the data fit perfectly to a straight line, and that one value gets bigger as the other gets bigger (like houses and price). An *r*-value of -1.0 means, again, a perfect straight-line fit, but one gets bigger as the other gets smaller, like cars and gas mileage. The correlations get closer to zero as the straight-line fit gets worse and worse, until there is no observable (linear) fit at all and the correlation is 0.0 exactly.

Of course, a straight-line fit is not necessarily the only kind of fit there could be. Figures 4.16 and 4.17 show examples where there is a strong relationship between two data sets, but Person's *r* detects no straight-line fit. Figure 4.16 is a graph of the so-called **absolute value function** (where you convert negative numbers to their corre-

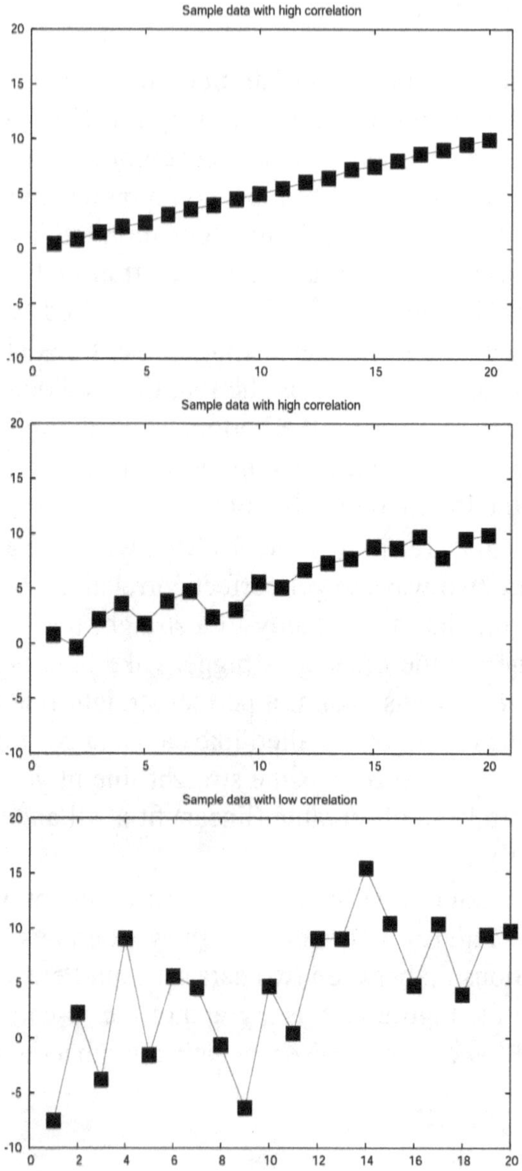

Figure 4.15: Examples of association

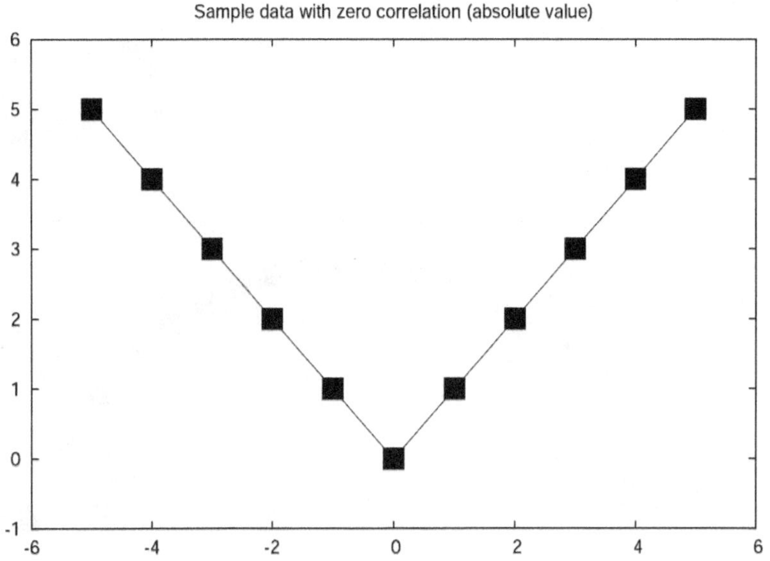

Figure 4.16: Example of a structured distribution with zero correlation

sponding positive numbers), but it could also be a measure of how prices change along a highway from west to east as you move away from the depot; figure 4.17 is a part of a simple circle, but could also be the pollution levels around that same depot (the closer you are, the higher the pollution). In either case, the lack of correlation means that there is no important difference between west and east. What is important is the simple distance, which is not directional.

Correlation can be very important for teasing out cause-and-effect

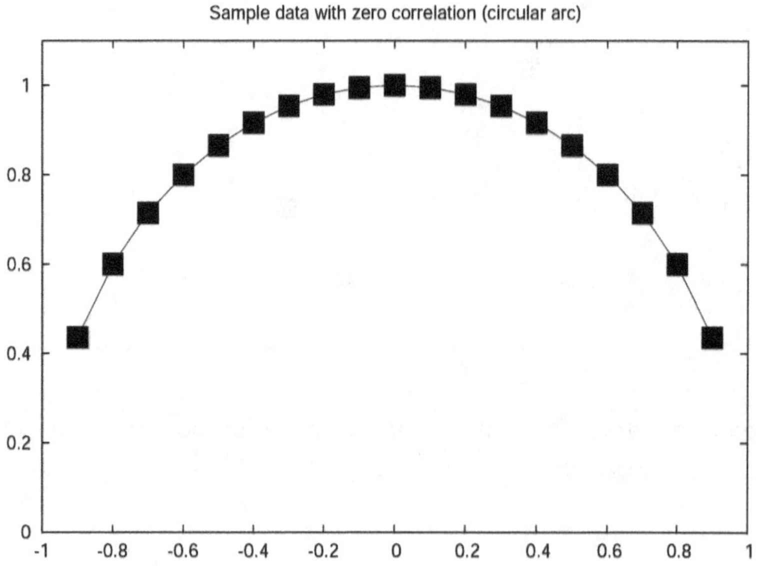

Figure 4.17: Another example of a structured distribution with zero correlation

relationships in data. For example, if there is a high correlation between a person's graduate education and income, then we can conclude that the two are related somehow. Either people with a high income are more likely to be able to afford grad school, or people with graduate degrees make more, or perhaps a third factor causes both (smart people both go to school and make more money).

Obviously, this must be approached with caution, because we don't know just from the correlation whether money causes schooling, schooling causes money, or whether some other factor causes both. Still, we can be confident that some relationship exists and we can use this to help us decide how best to look for it. It is true that some of the most truly breathtaking pseudoscience has been justified in the name of this caution (the correlation between smoking and cancer has been known for a long time, but *causal* data has been very hard to collect). So caution itself must also be approached with caution. What's the alternative idea, that cancer causes people to smoke? While that may be (barely) supportable mathematically, it lacks a foundation in common sense.

4.3 Hypothesis Testing

Using mathematics to test whether a plausible idea is likely to be true is one of the most important uses of statistics, and frankly, one of the hardest to get right. If all of Johnny's previous homeworks scored in the 30s, and then he suddenly scores in the 80s, did he cheat, or did he just luck out? Probing such questions takes us into the realm of **hypothesis testing.**

Unfortunately, statistics alone can't answer this question, and indeed understanding what questions statistics alone can and cannot answer is really at the heart of the problem. Statistics *can* tell us the likelihood that Johnny did something different—whether it be hiring a tutor, studying harder, or cheating—or whether Johnny simply got lucky this week and this is part of a normal variation in his scores. But while statistics may be able to tell you that blondes make more money on average, they can't tell you whether that's *because* people pay blondes more, or *because* people who make a lot of money are more likely to dye their hair blonde. This sounds like an obvious distinction, but it's a subtle one in practice. Often, it is precisely the cause that we're after, and statistical information can help us toward conclusions about where causes lie. But that is not at all the same thing as saying that the statistics establish or prove a causal link. "Correlation is not causality" is something like a proverb among people who work with statistical data. The Roman rhetoricians spoke of it as a particularly vicious fallacy, naming it *"cum hoc ergo propter hoc"* ("with this, therefore because of this.")

4.3.1 Null and Experimental Hypotheses

All hypothesis testing is done against a background of what is called the **null hypothesis**: the idea that nothing interesting is going on and that only random variation is occuring. For example, if we are interested in the idea that men have worse handwriting than women, the corresponding null hypothesis would be that men and women have, on average, equally bad handwriting. This null hypothesis is usually referred to as H_0. H_1, by contrast, is the **experimental**

hypothesis; in this case, the idea that men's handwriting is worse (or at least different) from women's.

Unfortunately, it's not practical to look directly for evidence that men's and women's handwriting differs. This might seem odd, but what happens if you look for evidence and don't find it? Does that mean that there is no evidence, or does it mean that you didn't look hard enough or in the right place? If the differences are real, but subtle, then we don't want to say "there is no difference" when all we really know is that we didn't find any. So instead, statisticians approach the problem in reverse, by trying to find evidence that the null hypothesis is not true.

There are, therefore, two possible outcomes of any hypothesis test. One can either *reject* the null hypothesis (if we find enough evidence to convince us that it is not true), or else simply *fail to reject* the null hypothesis, in which case we have, formally, learned nothing.[7]

The main reason that we do it this way, however, is that for a sensibly chosen null hypothesis, we can calculate (numerically) how unlikely a given outcome is based on the typical patterns that we see. For example, it is common knowledge that an average house cat—say, a Siamese—weighs about ten pounds. If you visit someone's house and see a ten-pound cat with Siamese markings, you accept that as consonant with reality. But if you visit and see a forty-pound

[7]Of course, informally, if we fail-to-reject often enough, we will often conclude that the null hypothesis is true. Patrick can't prove that unicorns don't exist, but if he looks for them often enough and doesn't find them, that might be good enough to satisfy him. On the other hand, that never seems to satisfy Steve.

Out of 100 Siamese cats measured,	weighed this much or more	Percentage (or likelihood)
85	5	85%
50	10	50%
30	15	30%
15	20	15%
7	25	7%
2	30	2%
0	40	$< 1\%$

Table 4.6: Fictitious weights of 100 surveyed Siamese cats

monster, even with the Siamese markings, you would be much more likely to reject the idea that this is an ordinary Siamese, in favor of the idea that it was some sort of unusual breed that also had color-point markings.

Where, though, do you draw the line? Would you reject a thirty-pound monster? Twenty-five? Fifteen? This is where the statistics come in. Consider the (completely fictitious) data given in table 4.6. Out of 100 Siamese cats measured, 50 weigh at least 10 pounds. Generalizing this to the population at large, it implies that if you pick up a cat at random, it has a 50/50 chance of weighing at least ten pounds. Only 15 cats weighed more than 20 pounds, so a random Siamese would be expected to have about a 15% chance ($p = 0.15$) of weighing 20 or more pounds. Only 7 cats weighed 25 pounds or more, and no cat measured weighed forty or more.

This table, then, gives us an estimate of the probability that if the cat sitting in front of you were Siamese, it would weigh as much as it does (or more). The technical term for this estimate is the **probability value** or more simply the ***p*-value**—the probability that the observed data would be seen if the null hypothesis were true. At this point, the test becomes simple logic. If the cat were an ordinary Siamese, it would probably not weigh forty pounds. Therefore, if it does weigh forty pounds, it's probably not an ordinary Siamese. The number from the table gives us an exact figure for "probably."

4.3.2 Alpha and Beta Cutoffs

If the probability that we get our observed data given the null hypothesis (i.e., the likelihood of a normal Siamese being that heavy) is lower than some threshold, we reject the null hypothesis as being incompatible with our observations. We could say, for example, that if the chance of something happening is less than 5% (thus setting the threshold to 5%), we reject it. In this case, we would accept the idea of a twenty-five pound Siamese, but reject the idea of a thirty-pound one.

Choosing the threshold can be tricky, because there are two potential errors that need to be balanced against each other. On the one hand, you could overzealously reject the null hypothesis when it happens to be true (in which case, Fluffy really needs to go on a diet!). On the other hand, you could naively accept the null hypothesis ("How dare you think that I would let my cat get that unhealthily obese! She's not a Siamese, she's half Maine Coon."). In either case, there are possible negative consequences—either you are putting the cat's health at risk, or you risk insulting a friend. Which is a more important error for you to avoid?

Our example is a bit fanciful, but serious examples abound. If the numbers for a particular disease (say, tuberculosis or influenza) are up from last year, that might just be ordinary year-to-year variation. It might, however, indicate the emergence of a new strain, and thus be the leading edge of a major health emergency. Ignoring a genuine problem might put much of the population at risk, but taking unneeded drastic measures will cost time, money, and might cause panic and other social problems (along with the loss of confidence when it turns out to be unneeded).[8]

The formal term for the rejection threshold is the **alpha cutoff**, and it's usually described with the Greek letter α. If your measured probability value is less than α, you reject the null hypothesis. Essentially, we are dividing the world into "common" and "uncommon" events, with a clear, bright line at α, and saying that we don't believe that "uncommon" events happen by chance.

By convention, most science and social science papers will use an α cutoff of 0.05, meaning that "statistically significant" results are results that were less likely to have appeared by chance alone than one time in twenty. Closely tied to this is the **beta cutoff** (β), which is the probability that you missed something that was really there by accepting the null hypothesis. Unlike α, β cannot be easily manipulated, as it depends not just on α, but also on the kind of data you have, how much of it you have, and what you plan to do with it.

The easiest way to adjust β is to adjust α—reducing the chance

[8]The SARS outbreak n 2003—which cost the city of Toronto hundreds of millions of dollars—provides a recent and poignant example of this.

of rejection will automatically increase the chance of acceptance, including an erroneous acceptance. If the consequences of one type of error are significantly worse than another type (William Blackstone famously thought that "it is better to let ten guilty men go free than to let one innocent be punished"), then the alpha cutoff can be set particularly high or low as appropriate. To reduce both β and α at the same time is more difficult; you typically can only do that by gathering more data, which may be difficult or impossible. Finally, depending upon what else you know about the data (or are willing to assume that you know), it may be possible to perform more accurate tests, as we will see in the following sections.

An easy way to test the hypothesis that there is some relationship between handwriting and gender would be to gather a pool of people, say a first-year college class, and ask them to write a paragraph or so. I then have an independent rater judge how legible each paragraph is, on a scale from 10 (perfectly clear) to 1 (this looks like a prescription pad). Our experimental hypothesis is that men will be less legible, on average, than women. H_0 is that women and men will be equally legible.

Our thirty subjects produce the data listed in table 4.7. Calculating the averages (means, and equally importantly, standard deviations), we see that men are in fact slightly less legible on average. But is this a real difference? Would I have gotten an equally large difference the other way in a different classroom? To determine that, we need to quantify the likelihood that we would have gotten this big a difference, even if H_0 were true. This brings us to the subject of particular methods for hypothesis testing.

Subject number	Man	Woman
# 1	1	2
# 2	2	7
# 3	3	5
# 4	4	4
# 5	5	9
# 6	6	3
# 7	7	8
# 8	8	10
# 9	9	6
# 10	2	7
# 11	4	2
# 12	3	9
# 13	9	8
# 14	8	3
Mean	5.0714	5.9286
SD	2.7586	2.7586

Table 4.7: Legibility judgments for the handwriting of 15 men and women

4.3.3 Parametric Tests

There are two general types of statistical tests: **parametric tests** and **nonparametric tests**. Parametric tests are those that make assumptions about the underlying shape of the data. For example, a normal (bell) curve is symmetric—indicating, for example, that there are as many people 50% above the average as there are 50% below the average. If your experimental data is bell-shaped, you can carry this theoretical assumption out into the real world. But in many cases, the data will not be symmetrical, and you will make an error if you assume such a high degree of symmetry. The average weight of a cat may be ten pounds, but there are more cats that weigh 25 pounds, fifteen pounds above the average, than weigh -5 pounds, fifteen pounds below. Despite these limitations, parametric tests are probably the most commonly used tests, because the assumptions are generally easy to conceive, and are often met in the real world.

Tests can be further subdivided by the type of question they purport to answer. For example, "Is this (individual) cat too heavy to be a Siamese?" is a question of whether a single data point is likely to have come from a known data set. "Is men's handwriting less legible on average than women's?" is a question about whether two different data sets have the same mean. "Do intelligent people have larger brains?" is a question about the relationship between two distributions, in effect asking whether cases that are high on one scale (tested intelligence) are also likely to score highly on another scale (brain volume).

Space prevents us from giving detailed descriptions of all of the tests that have been proposed, but we will try to explain enough that you can understand some of the major ones. The most important thing is to try to understand the assumptions that underlie the tests, because those are what determine whether the use of a particular test in a particular case is justified.

Comparing a Data Point to a Group with Known Characteristics.

This simplest of all statistical tests is designed to determine how likely an individual is to have come from a known population (as with our forty-pound monster cat above). In this case, if the population is known to have a normal distribution, with a known mean and standard deviation, we can simply calculate how far (in terms of standard deviations) the data point is from the mean. If our cats have, for instance, a mean of 10 and a standard deviation of 5, a twenty-pound cat is two standard deviations above the norm. A forty-pound cat is six standard deviations above the norm.

The relevant probability can be read directly from the charts describing the standard normal distribution. For example, approximately 14% of the population will be one or more standard deviations above the norm. Only 2.2% will be two or more standard deviations above the norm. Less than one half of 1% of the population will be three or more standard deviations above the norm.

For historical reasons, the standard normal distribution is sometimes called the ***z*-distribution,** and for this reason, this test is called the ***z*-test.** As a parametric test, it is only appropriate when the specific assumptions are legitimate. So in this case, to use a *z*-test, we need to know (or assume) that the population at large follows a normal distribution, that we know the mean and the standard deviations for the population at large (not just for our sample), that the sample is big enough (thirty to fifty data points is a good rule of thumb), and that the population at large is sufficiently larger than the sample (5 or 10 times larger, at least).

When the conditions are met, this kind of test can also be used to compare a single data point to several groups, in an effort to classify it (to determine into which group it should be placed). For example, if you would expect 8% of group A to score at your data point, but only 0.1% of group B to score that extremely, then you can say with relative confidence that your data point is more likely to come from group A.

Comparing a Group to Another of Known Characteristics

The z-test can also be used to compare one group against another group whose descriptive statistics are known exactly. For example, we mentioned earlier that IQ tests are designed to have results that are normally distributed, with a mean of 100 and a standard deviation of 15. To test whether college students have higher IQs than the population at large, you could simply get a large enough and representative enough sample of college students, give them IQ tests, and then use a z-test to see how far from 100 the sample average is.

Comparing Two Groups

The next simplest type of test is the one where you compare two groups. You can either do this directly (is experimental group A different from experimental group B?), or you can compare one group against a "known" background (is experimental group A different from the general population whose characteristics I can look up?). This is slightly more difficult than the preceding case because you have an additional source of error to contend with; the data of your group(s) are almost certainly not exactly what the "true" average should be, for the same reason that if you flip fifty coins, you're not likely to get exactly twenty-five heads.

For example,[9] one researcher used a standard psychological test (the Rosenberg test of self-esteem) to measure the self-esteem of a group of 112 Wellesley undergraduates. The published norms for this test state that the average score for the general public is 3.9. The Wellesley undergraduates scored an average of 4.04—slightly above average—but with a group standard deviation of 0.6542, reflecting the fact that individual scores can, and do, vary all over the map. The question was then posed whether these data show that Wellesley students actually have a higher average self-esteem than the population at large. A related but different example would arise if the same researcher had given a self-esteem test to a group of Harvard undergraduates as well, gotten a separate mean (and standard deviation) for Harvard, and then asked whether Harvard students really differed from Wellesley students in their self-esteem.

The most common parametric test for this situation is the ***t*-test,** which comes in a variety of subtypes (such as paired *t*-tests or pooled *t*-tests). This test is appropriate when all groups involved are nor-

[9]See `http://www.wellesley.edu/Psychology/Psych205/onettest.html`

mally distributed and the individual sample standard deviations are known, but the deviations of the population-at-large is not. (This is probably the most common situation one encounters, which is why the t-test is so widely used). When these assumptions are met, one can calculate from the means, the standard deviations, and the size of each group a probability that both groups were "really" drawn from the same underlying group (more technically, from groups with the same underlying mean, since Harvard students are obviously a different group of people than Wellesley students). In other words, we can assess the likelihood that Wellesley students, Harvard students, and the public at large are alike with regard to their (mean) self-esteem. If this probability is low enough, the null hypothesis will be rejected (as in the study cited).

So what's the difference between this study and the IQ study in the previous paragraph? With the IQ study, we know the mean and the standard deviation of the general population—not only what the average score is, but also how individual people or small groups are expected to vary from the average. With the self-esteem study, we know the average, but not the variation around the average. We therefore need to estimate this variation, which is built into the calculations for the t-test.

Comparing Three or More Groups

The t-test is not appropriate when one is comparing three or more groups simultaneously. For example, if you wanted to know whether all college students in the Boston area had the same self-esteem, you could gather groups from Harvard, MIT, Wellesley, BU, BC, Northeastern, Tufts, and UMass. This group of eight schools would yield 28 separate pairings between schools that you could test using *t*-tests. However, this would raise the chance of a false rejection error to an unacceptably high level (almost to the level of a dead certainty, in fact). (Remember that each test has its own chance of a false-rejection error, typically set at 0.05 or one in twenty; taking twenty-eight shots at a basketball hoop at one chance in twenty makes it very likely to hit.)

Instead, we want to compare all the groups at once. The appropriate parametric test to use in this instance is the **ANOVA** (short for "analysis of variance") test, which is essentially a generalization of the *t*-test that controls for the multiple shots at the target problem. Again, this is a parametric test, and only useful when the underlying groups are normally distributed. Furthermore, the variance must be the same across groups. Like the *t*-test, there are several different types of ANOVA depending upon the exact question you are asking.

Measuring Association

The measurements of association discussed in section 4.2.5 give direct rise to a test for association. If you want to know, for example, if there is a relationship between brain size and intelligence, measure both and determine what the association is. The standard methods for measuring correlations (Pearson's *r)* can be turned directly into a probability measure when the size of the samples are known (and such tables are widely available). The same cautions apply to this test as to any others, though. Pearson's *r* describes only how well the data can be fit to a straight line; data that fits perfectly to a curved line may have a low straight-line correlation when measured.

More technically, the calculations of Pearson's r make some assumptions about the underlying data. In particular, it assumes that the "residuals" (how far each point is from the regression line) are normally distributed, and that the variability of the residuals are the same regardless of the size of the individual data points. In some cases, these assumptions are clearly false. For example, there is a clear relationship between age and height among children, but the variability in height is much greater for older children than for younger ones. (A three year old who is a foot taller than average is a giant. A fourteen year old who is a foot taller than average is merely a basketball player.) If these assumptions are not met, then *hic sunt dracones.*

To return to the handwriting legibility study: We are comparing two groups, so a t-test is (or might be) appropriate. Running that test, we get a p-value of about 0.42, meaning that even if men and women were exactly the same in terms of average legibility, we'd see that big a difference more than 40% of the time. Based on this, we "fail to reject the null hypothesis" and we conclude that we have no reason to believe that men and women differ in handwriting legibility.

Implementing Tests

Perhaps it is a good time for us to mention that few people do any of these calculations by hand. In many cases, the mathematics is easy enough, but rather tedious to compute manually (ANOVA, in particular, involves some quite involved calculations). For this reason, we (and nearly everyone else who works with statistical data) use statistical software packages to analyze data. There are many such packages available,[10] and all of them are capable not only of conducting nearly any conceivable statistical test, but have facilities for manipulating all manner of data and generating plots and graphs that take advantage of established techniques in data visualization. Even ordinary spreadsheet software usually comes with a vast battery of statistical functions.

Often, these packages allow you to pull "t-test" out of a drop-down menu or simply type a command. The important thing, of course, is to know whether the test is appropriate or not. We have both seen research results that seemed to us to be sustained mainly by overeager (and ill-informed) exploitation of the capabilities of the software. On the other hand, neither of us would consent to analyze data without such programs, and you can usually be confident that the implementation of the tests and analysis routines have been pored over and tested for accuracy.

4.3.4 Non-parametric Tests

So what do we do when we meet dragons (situations in which we cannot make assumptions about the underlying shape of the data)? The simplest answer is just to cross our fingers and use the parametric tests anyway; in many cases, the assumptions, while formally necessary, can be violated without damaging the accuracy of the test or the inferences drawn from it. Looking at the sensitivity of various standard tests to different kinds of assumption-violations is a sub-

[10] Our favorite, a package called R, is free (`http://www.r-project.org/`)

stantial research area in modern statistics. In particular, the aforementioned Central Limit Theorem states that if you sample often enough from an oddly-shaped distribution, the sum (or the average) of your samples will be normally distributed. In many cases, this means that if you simply have enough data—some have proposed, as a general rule of thumb, two dozen or so data points–it's safe to use parametric tests.

But what if you don't have enough data? In such cases, statisticians have proposed **non-parametric tests**: tests that do not make assumptions about the underlying distribution. Each of the standard tests discussed above has one or several nonparametric equivalents. For example, instead of using the t-test, you can use the **Wilcoxon** test or **Mann-Whitney** test. Instead of the ANOVA, use the **Kruskal-Wallis** or **Friedman** test. For Pearson's r, use **Spearman's** ρ. The details of these can be found in any statistics text, and once again, the statistics packages know all about them.

The main problem with nonparametric tests is that they tend to be weaker (in the sense of beta cutoffs) than their parametric cousins. For example, the Wilcoxon test will sometimes fail to reject the null hypothesis where a standard t-test would reject it on the same (valid) dataset. The only way to address this is to provide more data, and studies suggest that a t-test only needs about 95% of the data that a Wilcoxon test needs, or, alternatively, that a Wilcoxon test can only pull 95% of the information out of the data that the t-test can. Similarly, Spearman's ρ has about 91% of the power of Pearson's r, which means that if you need 100 data points to find an effect using Spearman's test, you would only need 91 to use Pearson's test (assuming it's appropriate).

What kind of test should you use? There are really four cases. Either you have lots and lots of data—enough to overcome the probability that the test will reject the null hypothesis when the null hypothesis is actually false—or you don't, and you need to be cautious. Likewise, you either have data that fits a nice "normal" shape, or you don't. If you've got lots of normally-distributed data, then a parametric test is appropriate and more efficient than a nonparametric one, but 91% or 95% of lots of information is still lots of information. If you have lots of non-normal data, then the Central Limit Theorem tells us that it's probably safe to use the parametric test anyway. So in either case, the problems caused by using the wrong type of test are not that serious.

It's when you get into small data sets that the problems arise. Using a non-parametric test on normally-distributed data will lose power and efficiency, costing information that you probably didn't want to lose. The calculated probability values are likely to be high, and so you are more likely to make a false acceptance of the null hypothesis. If you use a parametric test inappropriately on small data, the results could be almost anything. The probability value is likely to be inaccurate—and you wouldn't even know whether it's more likely to be overstated or understated.

So how does one avoid these problems? There are two ways. The first is simply to double up—to run non-parametric tests in conjunction with any parametric tests, and look for differences (a matter made trivial by computer software).[11] The second is to examine the

[11] Patrick is a big fan of non-parametric statistics, and uses them professionally at almost every occasion, either because he doesn't trust the underlying assump-

data themselves prior to analysis, to see if the data are actually distributed normally. This can be done easily just by eyeballing it. For example, here is the distribution of the data for men's handwriting legibility (presented earlier as table 4.7):

Legibility rating	# instances
1	1
2	2
3	2
4	2
5	1
6	1
7	1
8	2
9	2
10	1

Does this look bell-shaped to you? Not to us, it doesn't.[12] Since the distribution is so far from bell-shaped, and we've only got about 15 data points, we have to use a nonparametric test in this case. Since what we were performing is a comparison between two groups, the

tions, or just as a reality-check on the results he gets with parametric stats. It's always nice to have a backup. If Pearson says "yes" but Spearman says "no," for example, then he assumes that something unusual is going on and that he should look more closely at the data to see just what it is and whether it can be fixed. When the same thing happens to Steve, he goes off and writes an article about postmodern indeterminacy.

[12]There are even some formal statistical tests, such as the **Kolmogorov-Smirnoff** and **Shapiro-Wilkins** tests, that will tell you whether a distribution is close enough to bell-shaped. We won't go into them here, but again, a good statistics textbook will treat of the subject in full.

appropriate test would be the Mann-Whitney test.[13]

4.3.5 Interpreting Test Results

As we have seen, the interpretation of test results can be tricky. Superficially, one ordinarily compares the observed probability value to the previously decided alpha cutoff, and decides whether to reject the null hypothesis.

But what does this mean, exactly? If the null hypothesis is rejected, then we conclude that the experimental hypothesis is true. If the null hypothesis is not rejected, then we conclude that there is not enough evidence in this test to decide one way or another. Notice that in this framework, we can never conclude that the experimental hypothesis is false. This is one of the meanings of the oft-cited maxim that "you can't prove a negative." Of course, you *can* prove a negative. Neither of us have tigers in our refrigerators; we just checked. You can't, however, prove a negative *by statistical analysis.* I can't prove, for example, that drinking grape Kool-Aid will not cure cancer. It might have some effect that's too small for me to have seen in the experiments I did. Maybe if I had studied fifty thousand people instead of fifty, something would have shown up. Maybe if I had run my unicorn detectors for a hundred years instead

[13]The reader might wonder where on earth we got data on handwriting legibility with this kind of odd distribution, since most data sets from this kind of social science analysis are much more "normal." A really suspicious reader might think we made the numbers up out of thin air. Truth is, we did. We have no idea about comparative legibility of men's and women's handwriting. In fact, we're not even sure we remember how to write by hand.

of a week, I would have found one.

Or not. But you can't use statistics to prove I wouldn't have.

Interpretation of the probability values is an issue as well. Technically speaking, if we reject the null hypothesis (say we got a probability value of 0.02, and we used an alpha cutoff of 0.05), we still run the risk of making a mistake. No statistical conclusions can be drawn with 100% confidence. But does this mean that (since we got a 0.02), that our conclusion is 98% likely to be true?

Unfortunately not. What if these are the results coming out of our unicorn detector? Our value only says that *if* the null hypothesis were true, then there would be only a probability of 0.02 (2%) of getting the results we did. But if we run our unicorn detector a hundred times, we would have expected to see such a result twice. Knowing as you do that unicorns don't exist (sorry, Steve), you will have no problem concluding that we just got lucky—or unlucky, depending upon your point of view. There's still a 0% chance that we detected a unicorn, and a 100% chance that something unlikely happened to happen.

This is the final point in interpreting test results and statistics, and we hope you'll pardon us for stating the obvious: Use common sense. Spurious findings can and do occur (indeed, if you do twenty experiments, the odds are in favor of you finding at least one spurious result at the 0.05 level). For a brilliant illustration of this (with, unfortunately, an aspect ratio that won't fit in this book) we recommend you to the XKCD web comic and specifically to `https://xkcd.com/882/`. If a result is simply nonsensical, then treat it with appropriate caution. As an illustration of this, we would like to share a true, genuine story of an actual event, with only a few details changed to minimize the embarrassment of all concerned.

At a conference in Scandinavia years ago, a European researcher was presenting a paper on some sort of social behavior: let's say it was the distribution of drink purchases at a German bar, or the distribution of glottal stops in Czech place names, or something like that. The researcher's conclusion, proudly trumpeted through the medium of PowerPoint, was that the distribution was best described by a "confluent hypergeometric distribution." At this point, the person sitting next to Patrick, a very gifted lecturer in statistics at a major British university, leaned over and whispered in his ear, "Now, there's a man with a statistics program too powerful for him."

Why did she (the statistics professor) think that? Well, the results violated common sense. Don't worry about not knowing the meaning of a "confluent hypergeometic distribution." No one does—not even the statistics Ph.D.'s that we've asked about it. Some of them—a relative few— do know what a "hypergeometric distribution" is; it's a special kind of non-normal distribution that you see about once a decade, under very restrictive circumstances. It's related to the idea of product testing [the number of "successes" (products that aren't broken] that you get when you take products in succession out of a box without putting them back after testing.

The key insight here is that this has nothing to do with either drink ordering or glottal stops; in particular, the key aspect "without putting them back" means that once you've tested a particular widget, it's no longer available to be tested. (The technical term for this is "without replacement.") But, of course, if I decide to pronounce my village name with a glottal stop, that doesn't make glottal stops unavailable. It's not like there's a world supply of glottal stops that will be exhausted. And although there is a world supply of whisky, it's not going to be exhausted in a single night in a single bar. The underlying assumptions that give rise to the hypergeometric distribution simply don't fit the environment in which the researcher was working.

On the other hand, a test for a confluent hypergeometric distribution—a parametric test with very strong assumptions built in—is apparently distributed as part of one major statistical package (the lecturer could even tell me which one). The researcher had apparently gone one-by-one through all the tests in the drop-down menus looking for the one with the smallest p-value. If there are fifty tests in the package, he was almost certain to find one he liked. He triumphantly dragged it out, without thinking about whether it made sense, or if it was really true. To the amusement of the professional statisticians in the audience, it almost certainly wasn't. In a sense, he was lying to himself, at least by omission, by not checking to see whether his story was plausible.

4.4 Some Humanistic Examples

4.4.1 Hemingway's Sentences

Statistics works out to be one of the most important tools in the workshop of the mathematically sophisticated humanist, precisely because (when properly done) it provides an independent voice in scholarly debate, allowing the data to speak for itself. It's a commonplace observation, for example, that the American novelist Ernest

Hemmingway wrote in a distinctive style, often described in terms like "short, choppy sentences." But in any sort of robust scholarly debate, one has to ask—how do you know this? And what does it mean? And, in particular, does Hemingway actually use "short" sentences relative to other writers?

The statistics, if you let the texts themselves narrate, tell a different story. The average sentence length in *The Sun Also Rises* is about 20.4 words. By contrast, the average sentence length of *The Great Gatsby* is 14.0, or in other words (in this sample), Hemingway writes sentences about half again *longer* than Fitzgerald. Amazon.com computes these figures for you, if you care, in the "Text Stats" sections of the web pages—a feature to which the authors find themselves entirely addicted. Perhaps more importantly, of the books which Amazon has studied, an astonishing 73% have a shorter sentence length. (This isn't an isolated finding. *The Old Man and the Sea* has an average sentence length of 16.0, and *For Whom The Bell Tolls* has 21.1.) The statistics tell a story of Hemingway, in fact, using sentences substantially longer than average.

Does this invalidate our intuitions that Hemingway "seems" terse, choppy, and simple? Of course not. But it does suggest that the apparent terseness comes from something other than just "short sentences." Instead of seeing this as a way of ending debate, we'd rather see it as opening the door to further discussion. If it's not short sentences, then what is it? How do we account for the variance between our impressions of the author's style and the empirical features of that style?

4.4.2 The *Federalist* Papers

Another story statistics can tell involves the tale of their authors. The emerging discipline of stylometry is one area where statistics can help to tell authors apart—an apparently straightforward matter that has deep implications for the study of style and influence. The analysis of the *The Federalist Papers* undertaken by Mosteller and Wallace is a classic example of Bayesian analysis that set the standard for much of the next fifty years work.[14]

The background is relatively straightforward: *The Federalist Papers* are a collection of eighteenth-century political essays written (anonymously) by some of the Founding Fathers of the United States, most notably Alexander Hamilton and James Madison. The question, of course, is which essays were written by which Father?

Mosteller and Wallace observed, in these authors' other writings, that they differed substantially in how they used various words, notably prepositions and conjunctions. For example, the following table looks at forty-eight Hamliton and fifty Madison papers and shows how the rate per 1000 words of the use of several words varies between Hamilton and Madison:

[14]Mosteller, Frederick and David L. Wallace. (2006). Inference and Disputed Authorship: *The Federalist*. Stanford: CSLI Publications. Reprint of 1962 edition.

Rate per 1000 words of	Hamilton *by*	Madison *by*	Hamilton *from*	Madison *from*
1–3	2		3	3
3–5	7		15	19
5–7	12	5	21	17
7–9	18	7	9	6
9–11	4	8		1
11–13	5	16		3
13–15		6		1
15–17		5		
17–19		3		

From this table, it should be apparent that the writing styles of Hamilton and Madison differ in statistically noticeable ways. Bayes' Theorem can then be used to help determine whether Madison or Hamilton is more likely to be the author of a given piece of text. For example, if we count 14 instances of the word "by" per thousand, we can follow more or less the same argument we used in section 4.1.3:

1. The probability of that particular document being written, which we will call $P(B)$, is what is is. As you will see, we don't even need to assess that.

2. Initially, we don't know which of Hamilton or Madison wrote the document: $P(H) = P(M) = 0.5$ (This is sometimes called the least informative prior because we assume nothing based on previous study.)

3. About 14/50 or 28% of Madison's writing uses *by* this often. $P(B|M) = 0.28$

4. Fewer than 1/48 (call it 2%) of Hamilton's writing uses *by* this often. $P(B|H) = 0.02$

   At this point, we're on familiar ground and among old friends:

5. Using Bayes's theorem, we find that

$$P(M|B) = \frac{P(B|M)P(M)}{P(B)} \qquad (4.36)$$

$$P(H|B) = \frac{P(B|H)P(H)}{P(B)} \qquad (4.37)$$

6. Now, we want to know which is more likely, $P(M|B)$ or $P(H|B)$. (In English, we want to know whether it's more likely that Madison or Hamilton would have written the document containing this particular word set.) So we can calculate the odds ratio by dividing one expression by the other (N.b. $P(H) = P(M)$ so we can divide both out.):

$$\frac{P(M|B)}{P(H|B)} = \frac{\frac{P(B|M)P(M)}{P(B)}}{\frac{P(B|H)P(H)}{P(B)}} \qquad (4.38)$$

$$= \frac{P(B|M)}{P(B|H)} \qquad (4.39)$$

$$= \frac{0.28}{0.02} \qquad (4.40)$$

$$= 14 \qquad (4.41)$$

We find, just by studying this one word, the the odds are approximately 14 times greater that Madison wrote the document than Hamilton. Of course, this is only one word among many. Most words don't produce this kind of separation (what do we do if the word "from" appears two per thousand words?) and we would ideally want to use lots of evidence. But the basic idea and analysis is sound, the argument is clear, and this method—Bayesian inference—can be widely applied to problems involving text analysis.

4.5 Lying: a How-To Guide

Despite the strength of this kind of analysis, there's still ultimately a strong reason why people distrust statistics: It's too easy to lie, and often too hard to catch.[15] One can lie by omission, simply by making mistakes or by not understanding, or one can lie more actively—but in either case, the result is less than illuminating.

The simplest type of lie is the simple mistake, using the wrong test or the wrong formula, or even making a mathematical error (though professional-grade statistics software makes that extremely unlikely). If you want to make your figures look good, just run through every test your software has to offer, and then tell your colleagues about the best results, making sure to pronounce the buzzwords clearly and confidently. More subtly, you can just confuse

[15] Years later, Patrick still hasn't received a satisfactory answer about what makes a hypergeometric distribution "confluent." No one seems to know. He may, however, owe the European researcher an apology when he finds out.

the mean, the median, and the mode; after all, the median income and the mean income in the millionaire's car are substantially different, but if you call them both "average," who will know which you meant?

If you're feeling assertive, you can simply force a false interpretation on your audience. If your α cutoff is 0.01, and you reject the null hypothesis, then that means that the chance of your experimental data having come about by chance is less than 1%. But why not simply tell them that your experimental hypothesis is 99% likely to be true? Sure, it's a lie, but it's much easier to understand, isn't it? Studies involving correlation are good candidates for this kind of forced interpretation; a correlation really says only that there is a relationship between the two groups studied. But it's easy enough to lie and claim you know what kind of relationship it is. If aspirin consumption is correlated with migraine headaches, then tell your audience that aspirin is dangerous because it causes migraines (and collect your pay from the anti-aspirin lobby). An observer paying close attention might realize that people who have migraines are more likely to take aspirin, so the causal link is likely to go the other way...but as long as you talk fast enough maybe the audience at large won't notice. It's true that, as the saying goes, "correlation (or association) does not imply causation," but you don't have to tell your audience that.

Similarly, you can leave out information. Telling us about the mean without telling us about the variation around the mean is a good way of leaving stuff out. For example, if you tell us that the average number of wheels per vehicle in a parking garage is 3.0, that conjures up visions of a building full of bizarre-looking tricycles.

Telling us that half have two and half have four informs me more honestly that a lot of the clients are into bikes.

A more serious way to leave out information is to distort it as it goes in, through selective reporting. A statistic is only as good as the data it's based on, and if you manage to avoid finding information that goes against your viewpoint, you don't have to use it. Darrell Huff has a great example in his book *How to Lie with Statistics* (Norton, 1954) about a report on the incomes of Yale graduates, collected by a survey twenty-five years after graduation—a survey, of course, of all the graduates whose incomes were known and who were willing to respond. Let's look at how this introduces two lies.

First, who are the people whose addresses we will know? Or to turn it around, who won't we be able to find? Certainly, we'll know where all the congressmen, the federal judges, and the ambassadors are. The ones we won't know will be the "unsuccessful" ones: the bartenders, taxi drivers, the unemployed, and anyone else who might look bad in the survey. Naturally, we don't want to include them—and we have a perfect excuse not to.

Secondly, what will people say when they respond? Will I respond honestly, or will I be tempted to shade the truth? And if so, will I want to make myself look good by raising my actual salary, or look bad by downplaying it? So again, we've provided an easy way to make sure that the data that comes in is "better than life." Naturally, this will make our survey data look much, much better for Yale.

This kind of data distortion is, as we all know, called "**bias**," and it's probably the most common way that people shade their statistics. Unfortunately, there's no real way to correct for it once the data has been collected and accepted; statistical techniques can analyze what you tell them, but they can't analyze what you carefully don't tell them.

A related problem is that of **representativeness**. A good way to lie is to measure something slightly different, but not different enough that the person notices what you did wrong. For example, we proposed earlier to study self-esteem among Boston-area college students by surveys at eight good, academically-oriented schools. But how well would the results transfer to other schools (either less competitive "second-tier" schools, or "art" schools such as Berklee College of Music, Longy, or MassArt)? Can we conclude anything about them? Bias introduced by unrepresentative data can be a real problem in psychology, because most psychology studies are done on students at the researcher's college. But are university students really representative of the general public? By and large, they are younger and of higher socioeconomic status. Perhaps they are different in other ways as well—and without testing the population at large, we have no way of knowing.

In this chapter we've tried to provide some armor against "lies, damned lies, and statistics," but we will continue to insist that the most important aspect of this is simply knowing which tests are appropriate in which circumstances and what the underlying concepts are. One doesn't need to know what a hypergeometric distribution is, or exactly how to calculate an ANOVA (though both matters can be illuminating for other reasons). One needs to know, instead, that the ANOVA is being properly applied (i.e. that the underlying data is the sort of thing we would expect to be normally distributed with common variance) and that a hypergeometric distribution isn't just a collection of buzzwords. One needs to know that a probability of 0.05 means "one time out of 20" or "nineteen to one against"—an event that is improbable but not unheard-of.

In the end, the damn lies of statistics are a lot like the damn lies of that other tool of great power known as rhetoric. And that is because statistics is ultimately a form of rhetoric. Whatever else we might do with statistics, we are always out to persuade others (or maybe just ourselves) that things are thus and so—usually as a way to change minds, and often as a way to inspire actions and inform decisions. Given that august role, it is perhaps more sensible that Socrates be executed for having made "the lesser argument seem the greater" that that we dismiss the whole matter as empty and insubstantial ("just rhetoric," "more damn lies"). There are truths, powerful truths, and "Aha" moments that only come when we run the numbers.

5 Analysis and Calculus

It is customary to portray calculus as a subject poorly taught, and to begin any discussion of it by inveighing against any number of failings on the part of teachers, students, textbook authors, and indeed, the entire edifice of mathematics education. But having now attempted to write a conceptual introduction to calculus, we find ourselves having great sympathy for anyone who would try to explain it. The problem, we think, has less to do with the difficulties of the subject itself, and more to do with a tension, which, though appearing in some potential form in most mathematical subjects, becomes particularly acute with calculus.

A "calculus" is a method or system of reasoning. So really, we've been putting forth various "calculi" right from the start. If the system we now propose to unfold has earned the right to be called *the* calculus, it's because it is one of the most eminently practical systems ever created for solving a truly vast set of problems. Calculus is, in this sense, the ultimate form of applied mathematics. We can think of very few general problems involving numbers that don't eventually wend their way around to the techniques of calculus—including most of the mathematics of interest to humanists.

But there's something else about calculus—something quite beyond any discussion involving the trajectory of rockets or the dynamics of population change—that endears mathematicians to this subject. Because at base, the problems calculus was designed to solve are full of riddles and paradoxes that take us deep into the nature of things like numbers, change, infinity, and time. In this sense, it is the ultimate form of pure mathematics. We can think of very few conceptual problems involving numbers that don't eventually wend their way around to the philosophical conundrums that gave rise the invention of the calculus, and which continue to inform its theoretical basis. It is, to put it plainly, one of the most elegant ideas in the history of thought.

The tension, then, is between practicality and theoretical musing: between solving problems and understanding the problem. Focus too much on the former, and the inner beauty of the thing is lost; focus too much on the latter, and the startling utility of the techniques is obscured. What's more, there are several different ways of thinking about the subject within this tension: engineers think about it differently from mathematicians, as physicists tend to think about it differently from statisticians. And we too have our own way of thinking about it. We are very pleased to be released from the impossible task of trying to negotiate this territory for a general course that might include all these constituencies, and entirely humbled by the efforts of those who do precisely that.

Our method will be to get at the concepts of calculus through the back door (in a way entirely innappropriate for an ordinary textbook) by speaking first about the loftier matters upon which its utility rests, and only then turning to the techniques that unfold from that musing. Our hope is to get at the Big Idea that is set in motion whenever the techniques of calculus are used.

Because the Ideas are very Big indeed.

5.1 The Mathematics of Change

The problem of change—how to describe it and predict its patterns—is, at root, a philosophical problem. Zeno of Elea (c. 490–c. 430 BCE), whose famous paradoxes are familiar even to grade school children, was among the first to give us a sense of the troubling complexities involved. One of his paradoxes, which appears in Aristotle's *Physics,* states, "[T]hat which is in locomotion must arrive at the half-way stage before it arrives at the goal" (VI:9,239b10). That is, if you are standing on one side of a room, and you want to get to the other side, you must perforce pass the halfway point between where you are and where you want to go. But there is, of course, a halfway point between *that* halfway point and where you stand. And there's another between those two points. Follow this through, and you may end up concluding that you can't get there. In fact, it would seem that you can't move at all, because when we express the halfway points as a series of fractions, we see that you must pass through an *infinite* number of halfway points:

$$\left\{\cdots\frac{1}{32},\frac{1}{16},\frac{1}{8},\frac{1}{4},\frac{1}{2},1\right\} \tag{5.1}$$

The missing ingredient, here, is time—a fact intuited by Aristotle, and worked out more fully in a blaze of insight by Archimedes (c. 287–c. 212 BCE). For it is obvious that while the halved distances are becoming infinitely smaller, the time taken to traverse them is also becoming infinitely minute. Or rather, it is not in the least bit obvious. Nearly two millenia would pass before anyone would truly feel that they had resolved the paradox, because a complete mathematical proof (of your reaching the other wall) is impossible without the contributions made by calculus to the understanding of infinity.

So before we plumb the depths of that understanding, we'd like to ensure that you are properly confused about infinity. That way, you'll be as baffled by the subject as mathematicians were prior to Newton, and therefore more likely to appreciate what he managed to do with his summer vacation in 1667.

5.1.1 Exhaustion

What is the area of a circle?

Well, it's πr^2, from secondary-school geometry. r is the radius of the circle and π is that irritating number 3.14159... that never ends (the computers, you may recall, blow up when Captain Kirk asks them for the final digit). And it's not really $\frac{22}{7}$, though that's a good guess. Eudoxos of Cnidus (c. 408–c. 347 BCE) is usually credited with that guess, and his method for arriving at it was, to the say the least, ingenious. It was also deeply unsatisfying.

The Greeks, whose mathematics mostly dwelt on problems in geometry, knew how to find the areas of polygons. They knew that the area of a square, for example, is the length of the side times itself. They also knew that the area of a triangle is one-half times the base times the height, so that an equilateral triangle is $\frac{s^2 \cdot \sqrt{3}}{4}$, where s is the length of the side. You probably don't remember the formula for

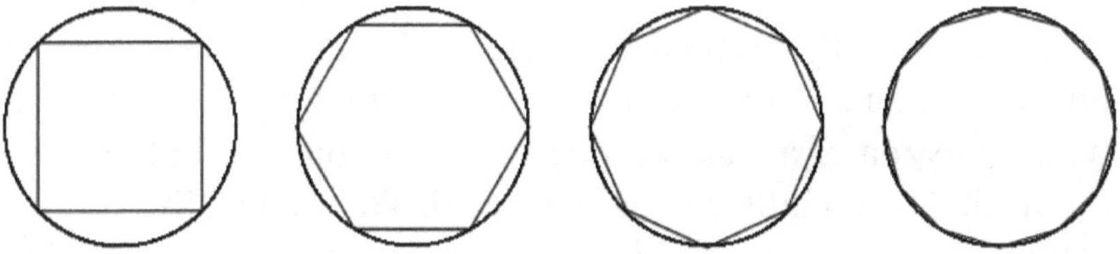

Figure 5.1: Exhausting the area of a circle via polygon (4-gon, 6-gon, 8-gon, 12-gon) approximations

the area of a regular pentagon. We don't remember it either, but the Greeks knew that one as well.

There's a big difference between these areas, though, and the area of a circle. A polygon has a few, straight sides, while a circle has a single, curved, side. Eudoxus' great insight was to use what he understood about polygons to get at the problem of things that didn't have any gons (the word "polygon" is a Greek neologism that means "many-angled.")

Consider the set of diagrams in figure 5.1. We've taken a circle and inscribed a number of polygons in it. Since the polygons are entirely inside the circle, the area of each individual polygon will be less than the area of the circle. But as we add more and more sides to the polygon, it appears that the polygon becomes closer and closer to the circle. Intuitively, we can suppose that as the number of sides gets larger and larger, we get a better and better approximation of the area of the circle.

The Greeks called this kind of analysis the **method of exhaus-**

tion. The idea is that the area outside the polygon (but inside the circle) is gradually reduced—"exhausted"—until there's none left and we have an exact value. But here's where things get confusing. Even a polygon of a hundred sides (or a million, or a billion sides) would still leave a little bit out in the cold. What we really need is to calculate what the area would be if we had as many sides as possible. But for that, we would need a number larger than all other possible numbers. An infinite number, in fact.

The concept of **infinity** has always exercised a powerful fascination on the imagination, particularly of science fiction writers ("To Infinity and Beyond!") Literally, it means "without bound" or "endless." The concept itself is not hard to imagine. What is hard to imagine, is dropping something "without bound" into a mathematical equation and treating as if it were just another number (like the number "2").

5.1.2 Paradoxes of Infinity

And that's because doing so seems to lead one into absurdities. A simple, non-numeric example is the "light bulb paradox." Some person has a light bulb and a switch, and is amusing themselves by flipping the switch on and off. If the person and the bulb are both immortal, what will the state be "at infinity?" Will the light be on or off?

A simple proof shows that the light will be off. After all, the light started out off (when the circuit was built), and every time the light is turned on, it is immediately turned off. Therefore, every time the light is turned on, it is turned off and it never stays on. So in the final state, it must be off.

However, an equally simple proof shows that the light will be on. After all, every time the light is turned off, it is immediately turned back on, and it therefore never stays off....

Nature, like mathematics, abhors a contradiction. The light cannot be both on and off. Therefore, there must be something wrong with one of the arguments. As it happens, there is the same thing wrong with both arguments: the idea that there is a "final state" (at "infinity") and we can just look at the final state to see what it is. The simple fact is that the "final state," if we can call it that, is of a switch being perpetually flipped without rest.

We can express this paradox in numeric form very easily. Consider the following (alternating) sequence:

1 + -1 + 1 + -1 + 1 + -1 + 1 + -1 + 1 ... *(and so on endlessly)*

The sum of the first five numbers is 1. The sum of the first six numbers is 0. But what is the sum of the infinitely long set of numbers in the full sequence?

Patrick argues: *"Well, I can re-write the sequence as*

(1 - 1) + (1 - 1) + (1 - 1) + (1 - 1) + (1 - 1) ... *(and so on endlessly)*

Since each of the quanitites in parentheses is equal to zero, this means the sequence as a whole is

0 + 0 + 0 + 0 + 0 ... *(and so on endlessly)*

so the final sum is 0."

Steve argues: *"You're quite wrong, as usual. The sequence is more properly rewritten as*

1 + (-1 + 1) + (-1 + 1) + (-1 + 1) . . . *(and so on endlessly)*

which is just

1 + 0 + 0 + 0 + 0 . . . *(and so on endlessly)*
which is just 1."

Of course, we're both wrong. This is just the light bulb paradox with numbers; the correct answer is that there is no "final sum." But here's the thing: there *are* infinitely long sums that *do* have final answers. Everyone will agree, for instance, that if you divide by 1 by 3, you get a repeating decimal 0.3333.... Using standard place values, this stands for the sum

3/10 + 3/100 + 3/1000 + 3/10000 . . . (and so on endlessly)

Since dividing 1 by 3 is equal to the value $\frac{1}{3}$, this endless sum must also have the value $\frac{1}{3}$ (what else would you get when you divide 1 by 3?). Similarly, three times this value, or 0.9999..., must have a well-defined final value of 1 itself. (What else could 3 times $\frac{1}{3}$ be?) So it must be possible to convert this endless sum into something tractable.

If "infinity" is not a number, then presumably we can't speak of it in the same terms we used in our chapter on algebra—which is to say, the operations we use on numbers are not well defined when speaking of "infinity." But the real problem is that this is only *sometimes* the case. So when does it make sense to talk about infinity and when doesn't it? When can we convert something "infinite" (like an endlessly repeating decimal) into something sensible like a number? As we said before, these questions have captured the imaginations (and sometimes the sanity) of some of the best mathematicians of the past four hundred years. And many of the answers are counterintuitive, if not outright strange.

To return to an earlier question: What does it mean to have an infinite number of anything? For example, if we claim that there are an infinite number of positive integers, how many is that? To understand this question, let's go back to sets for a moment.

Let's consider two sets: the set of positive integers $\{1,2,3,\ldots\}$ and the set of positive *even* integers $\{2,4,6,\ldots\}$. We'll call these two sets P and E for short. First of all, we claim that both of these are "infinite" sets, in the sense that if you started listing the elements of either set, you would never be able to finish. They are "endless," in an exact sense. But does this mean that they have the same number of elements? Well, on the one hand, every element of E can also be found in P, so there are at least as many elements in P as in E. We can therefore deduce that P is, at least, no smaller than E. But P also has some elements (in fact, an infinite number of elements) that are not found in E. If we were to take all the elements of E out of P, we'd still have some left. So this suggests that P is a *larger* "infinite set" than E (see figure 5.2).

Seems simple, yes?

Set P		Set E
1		
2	$\leftrightarrow$	2
3		
4	$\leftrightarrow$	4
5		
6	$\leftrightarrow$	6
7		
8	$\leftrightarrow$	8
	$\vdots$	

Figure 5.2: There are “obviously” more numbers than even numbers.

Set P		Set E
1	$\leftrightarrow$	2
2	$\leftrightarrow$	4
3	$\leftrightarrow$	6
4	$\leftrightarrow$	8
4	$\leftrightarrow$	10
5	$\leftrightarrow$	12
6	$\leftrightarrow$	14
7	$\leftrightarrow$	16
	$\vdots$	

Figure 5.3: But they are “obviously” matched 1-for-1.

But on the other hand, for every element p in P, there is exactly one element $2 \cdot p$ in E. (And for every element in E, there is exactly one element in P; just divide by 2. See figure 5.3.) So—just like "if you have one cup for every saucer and one saucer for every cup, you have the same number of cups and saucers" — this suggests that P and E are the same size and have the exact same number of elements.

Another paradox, and one that can only be resolved by thinking about what "infinite" really means. Again, the problem comes from trying to think about the end state of a process as though it were real (without justification). In this example, the mapping between the two sets is real and something we can create in the here-and-now, but the idea of taking all the elements of E out of P is itself an endless process. Yet this still leaves us with the seemingly paradoxical result that we can have two sets of the same size, with one being a subset of the other. And the only real resolution is to accept that our intuition may not be able to guide us properly in the ways of the infinite, and that the traditional rules of arithmetic that we learned in grade school may not apply.

May we blow your mind even further? Which is a bigger infinite set, the set of positive integers, or the set of fractions? The counterintuitive answer is that the two sets are exactly the same size, by roughly the same argument. We can set up a mapping between the set of positive integers and the set of number pairs by simply ordering the pairs by their total sum. For example, the set of positive integers begins {1,2,3,4,5,6,...}, while the set of number pairs begins with {(0,1),...} (the number pair that adds up to 1). It then continues with { ...(0,2), (1,1),...}, the pairs that add up to 2, and then {. . . (0,3), (1,2), (2,1) . . . }. Every positive integer corresponds to a (unique) pair.

It follows that every pair corresponds to a fraction : the pair (1,2) can be turned into $\frac{1}{2}$ without much effort. Similarly, the fraction $\frac{22}{7}$ corresponds to (22,7), which comes right between (21,8) and (23,6) in the list. Therefore, every positive integer corresponds to a fraction, and there are just as many positive integers as there are fractions.

Having said all that, the fact remains that *not all* infinite sets are the same size. For example, the set of all real-valued functions (this just means functions whose values are real numbers like 2.6 instead of, say, colors or animals) is *larger* than the set of all positive integers. The proof of this is a little complex, but very interesting, because the technique (**proof by contradiction**, more familiar to humanists as *reductio ad absurdam)* is commonly used in mathematics. So here, we're going to prove something to be true by rejecting the possibility of it being false. Specifically, we're going to follow an argument made by Georg Cantor (1845–1918), one of the first mathematicians to attempt to formalize the idea of the infinite.

We will start by assuming that the two sets are the same size, which is another way of saying that there is a mapping between all positive integers and all real-valued functions. (Because this is a proof by contradiction, we will eventually come back to this assumption and reject it.) We won't claim to know what that mapping looks like; perhaps function #1 (f_1) is the function $f_1(x) = x^2$ and function #5 is $f_5(x) = \frac{9}{5}x$. We'll simply assume that that there is some kind of mapping that relates *all* possible real-valued functions to *all* positive integers in a one-to-one relationship.

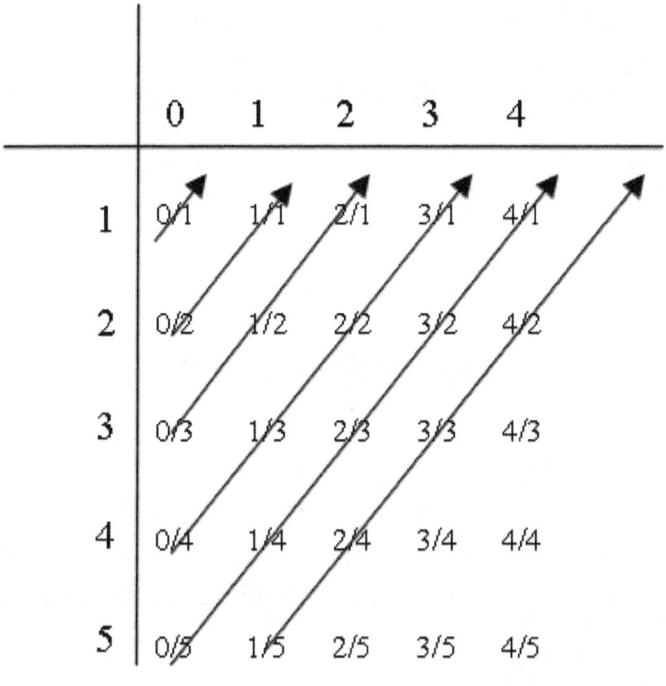

Figure 5.4: There is a 1:1 mapping between integers and fractions

Now, let's define a function D that takes each number and applies the function corresponding to that number to the number itself, and then adds 1. In other words, $D(1)$ equals $f_1(1)+1$, $D(5) = f_5(5)+1$, $D(75) = f_{75}(75)+1$, and in general $D(x) = f_x(x)+1$. We could call this a **diagonal** function, because when we arrange the correspondence between functions and integers as a table, we find that D is operating on the numbers that appear along diagonals of the table, like so:

Function	(1)	(2)	(3)	(4)	(5)	...
$f_1(x) = x^2$	***1***	4	9	16	25	...
$f_2(x) = x-1$	0	***1***	2	3	4	...
$f_3(x) = \frac{x^2}{2}$	1/2	2	***9/2***	8	25/2	...
$f_4(x) = -3$	-3	-3	-3	***-3***	-3	...
$f_5(x) = \frac{9}{5}x$	9/5	18/5	27/5	36/5	***9***	...
			⋮			
Diagonals $[f_x(x)]$	1	1	9/2	-3	9	...
$D(x) = f_x(x)+1$	2	2	11/2	-2	10	...

Now here's the question: if this list of functions really includes *all possible functions,* it must include D itself (since D is a function). But where?

D can't be function #1, since $D(1) = f_1(1)+1 > f_1(1)$. Since $D(1)$ and $f_1(1)$ are different, the functions D and f_1 are different. D can't be function #5, since $D(5) > f_5(5)$. D can't be f_{27} by a similar argument. In fact, D can't be anywhere in the list at all!

But this means that the *list of all possible real-valued functions* must be missing at least one function: namely, D. This is an obvious contradiction (our list that we assumed complete must be missing something), and therefore we know that our initial assumption, that such a list was possible, is false. Therefore, there is no mapping between integers and real-valued functions, and, although there are infinitely many positive integers, there are infinitely more functions. We therefore reach the unavoidable, if slightly uncomfortable conclusion that "infinity" comes in different sizes. We can therefore speak of larger and smaller "infinities."

The reaction to this work from Cantor's fellow mathematicians was not, shall we say, charitable. Like Gödel's theorem decades later, Cantor's ideas tended to overturn much thinking about mathematics itself. But while Gödel's theory caused a good deal of unease, it was nonetheless conceded to be correct. Cantor's ideas—which included, among other things, a proof that there were as many points on the plane formed by the unit square as there are on one of the lines that makes up the square's sides—were greeted in some quarters as a kind of heresy. Henri Poincaré (1854–1912) likened Cantor's ideas to a disease infecting mathematics; Ludwig Wittgenstein (1889–1951), who took a dim view of set theory in general, considered it laughable nonsense. The otherwise brilliant Leopold Kronecker (1823–1891) not only accused Cantor of being a charlatan and "corrupter of youth," but actively tried to supress the publication of his work. About the only scholars to welcome Cantor's ideas were the theologians.

Which is appropriate, because at heart this was a "religious" debate about mathematics—about its purity and integrity as a system and its presumed correspondence to the metaphysical fabric of the universe. In Kronecker's estimation, "God made integers; all else is the work of man" [2] To speak of "transfinite numbers" was to speak of things the angels dare not utter.

Cantor, sadly, went quietly mad, and most historians cannot avoid concluding that the repeated rejections and ridicule of his work by some of the most prominent mathematicians of the day had a hand in the progress of what was undoubtedly a quite severe case of mental illness.

There was one mathematician, though, who admired Cantor. In his *Autobiography,* Bertrand Russell—by then, Lord Russell—calmly averred that Cantor may well have been "one of the great intellects of the nineteenth century" [10].

5.2 Limits

But laying all of these objections aside, we are still left with the problem of having to deal with "infinity" in ways that do not involve mind-blowing levels of paradox. For reasons both practical and theoretical, we would like ordinary concepts of the infinite (repeating decimals like $0.3333\ldots$) to have unique, reasonable, sensible, and mathematically tractable meanings (as we do with things like $\frac{1}{3}$). If the entire concept of infinity is too difficult or paradoxical to work with, we need to find a paradox-free method of identifying concepts with which we can work using our ordinary methods and intuitions. The most common method involves the use of what are called **limits**.

5.2.1 Limits and Distance

The idea behind a limit is that of getting "close" to a given number or aspect. For example, in the "method of exhaustion" example given before, the successive polygons get closer and closer to an actual circle. Similarly, the area of the polygons gets closer and closer to the area of the circumscribed circle.

Let's formalize this a bit. Let A_c be the area of a circle, and let A_n be the area of a regular n-sided polygon inscribed in that circle. For example, A_4 would be the area of the square, while A_3 would be the area of the equilateral triangle and A_6 would be the area of the hexagon. (See figure 5.1 for pictures of A_4 and A_6.) We can then put forth the following two properties (which we could prove, but we will ask you to take our word for it).

$$\forall n > 2 : A_n < A_c \tag{5.2}$$

$$\forall n > 2 : A_n < A_{n+1} \tag{5.3}$$

1

Because these properties are true, we can use the estimates we are making of the area of the circle using finite polygons to establish a worst-case scenario—an upper bound on the error in our estimate.

[1]Recall from previous chapters that $\forall n$ means "For all n." In other words, for every value of n, the area of the n-sided polygon is smaller than the area of the circle, and also smaller than the area of the polygon with one more side. Why $n > 2$? Because there's no such thing as a 2-sided or 1-sided polygon, so A_2 doesn't makes sense.

At the very least, we know that our estimates will not get worse if we take more time and use more points in constructing our polygon. We could even establish an *a priori* limit on how bad an estimate we can accept for the area of the circle. Table 5.1 lists the areas for regular polygons of various sizes as a percentage of the circle covered; for example, a square ($n = 4$) covers about 64% of the circle it is inscribed in. If you need at least 90% coverage, you can use an 8-sided polygon—or, more precisely, any polygon with eight sides or more. If you need 95%, use fifteen sides or more. If you need 99%, use thirty or more.

As our approximation gets more and more accurate, we presumably get closer and closer to the actual area of the circle. On the other hand, we could keep adding "gons" forever. We could state this another way, though, and say that even though the process—the sequence of numbers representing the sides in this case—is infinite, there still must be some unique value toward which that process is tending. This unique value (if it exists) is the limit.

5.2.2 Formal Definitions of Limits

We just used the word "sequence" to describe the set of possible numbers of sides, and this term is useful for defining the concept of a limit more formally. A **sequence** is an ordered list of numbers. So if we have an infinite sequence of numbers A, we can refer to the individual numbers as a_1, a_2, a_3, and so forth.

We say that the sequence A has a limit L if and only if we can get arbitrarily close to L by taking larger and larger-numbered elements of the set. Say we want to get within 0.1 of L. If we can demonstrate

n	Area of n-gon of "radius" 1	Area of circle of radius 1	Percentage of circle covered
3	1.2990	3.141593	41.3497%
4	2.0000	3.141593	63.6620%
5	2.3776	3.141593	75.6827%
6	2.5981	3.141593	82.6993%
8	2.8284	3.141593	90.0316%
10	2.9389	3.141593	93.5489%
15	3.0505	3.141593	97.1012%
20	3.0902	3.141593	98.3632%
30	3.1187	3.141593	99.2705%
40	3.1287	3.141593	99.5893%
50	3.1333	3.141593	99.7370%
75	3.1379	3.141593	99.8831%
100	3.1395	3.141593	99.9342%
125	3.1403	3.141593	99.9579%

Table 5.1: Area of larger and larger polygons as a percentage of circle area

not only that any element of the sequence after, say, a_{64} is at least that close to L, but that this property holds *for any threshold we choose* (0.001 or 0.000001), then this sequence has a limit.

You want to get within 95% of the area of the circle? Use at least 15 points. You want 99%? Use at least 30. But the key thing is that however we define "close," the sequence both gets to and stays close to a particular, unique value.

Consider the example of the number 0.99999... (see figure 5.5). This, of course, is "really" an abbreviation for a sequence (0.9, 0.99, 0.999, 0.9999, ...) as we saw before. But we are now in a position to show that it has a well-defined limit value of 1; it both gets close to 1 (the number 0.99999 is within 0.0001 of 1), and stays close to 1 (every number after 0.99999 is also within 0.0001 of 1).

Or consider this sequence (which we will name F, for fraction, with the individual elements named $f_{something}$):

$f_1 = \frac{1}{1}, f_2 = \frac{1}{2}, f_3 = \frac{1}{3}, f_4 = \frac{1}{4}, \ldots$

We can formalize this sequence using the letter k as a general index variable—a placeholder for the kth element in the sequence. This lets us provide a general definition for the sequence elements by noting that $f_k = \frac{1}{k}$.

Obviously, this sequence gets smaller and smaller and will get very close to zero (without, as Zeno reminds us, actually getting there). But if I just want to be close to zero—say, within $\frac{1}{100}$ of it—then any element after f_{100} will do. If I want to be very close—say, within $\frac{1}{1{,}000{,}000}$—then any element after $f_{1{,}000{,}000}$ will do. You should be able to guess how many elements I would need to be within one part in a trillion of zero.

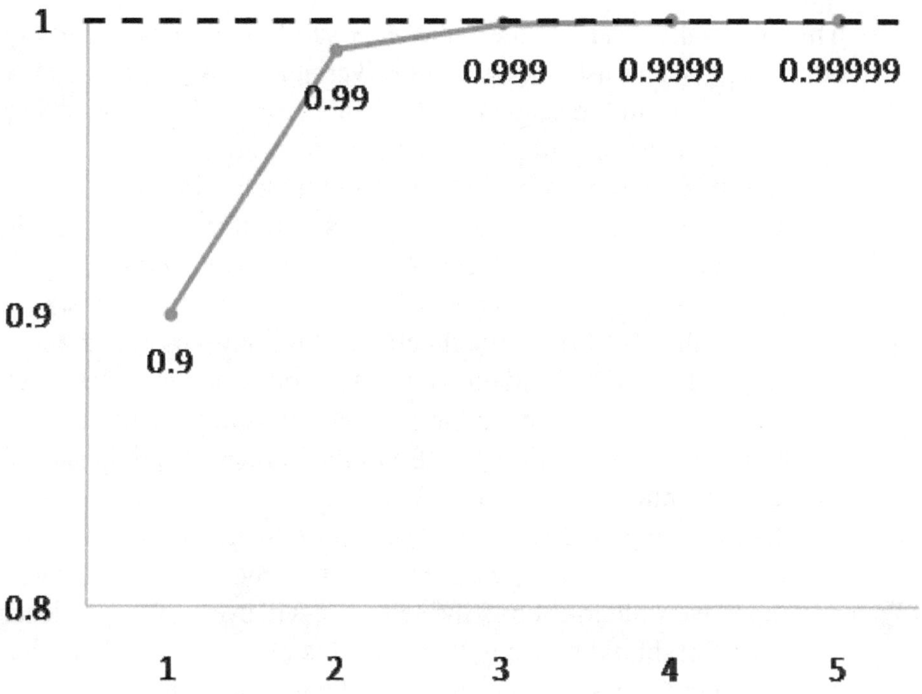

Figure 5.5: Limit of 0.99999... is 1.0 exactly.

The notation for this kind of limit looks like this: If our sequence A has a limit of L, we write:

$$\lim_{k \to \infty} A_k = L \tag{5.4}$$

This is usually read as "the limit of A sub k as k goes to infinity equals L."[2] We're using k as an index variable, so A_k represents the "k-th" element in the sequence. As k increases without bound, the value of the k-th element gets closer and closer to L.

Using this notation, we can formalize the definition of a limit as follows. Given a sequence, a limit exists only if there is a specific number that the sequence "gets close to" and "stays close to." To make things slighly easier on ourselves, we'll describe that distance in terms of the **absolute value** function. This allows us to talk about the "size" of a number without worrying about whether it's positive or negative (so both +3 and -3 have the same absolute value: 3). The absolute value of x is written $|x|$. So in this framework, the distance between two numbers x and y is simply $|x - y|$.)

When we say that a sequence "gets close to" a number, what do we mean? In this context, we mean that for any arbitrary threshold of "close" we want to choose, the sequence will eventually get closer than that threshhold and stay there. In this example, we claim that the sequence F has a limit of zero (which, again, means that the sequence F gets and stays close to zero.) We may be able to illustrate this by looking at some some numbers that don't have this property.

[2] This is the first time we've seen the ∞ symbol (sometimes called a **lemniscate**). This is the conventional mathematical symbol for infinity, but it's important to remember that it represents a concept (not a number of any kind).

For example, the sequence F never "gets close" to 8—in fact, it never gets within 7, let alone 0.1, of 8. The sequence does "gets close" to 1/2 (in fact, f_2 is 1/2 exactly), but it doesn't *stay close*. It does, however, both get and stay close to 0. Again, we can formalize this by noting that, for any value of ε—the conventional Greek letter used to denote an arbitrarily small, positive quantity—the distance $|f_n - 0|$ is less than ε if and only if $n > \frac{1}{\varepsilon}$. We can then say that the limit of A exists and is equal to L when the following holds:

$$\lim_{k \to +\infty} A_k = L \leftrightarrow \forall \varepsilon > 0 : \exists n : \forall m \geq n : |A_m - L| < \varepsilon \tag{5.5}$$

Now, this may well be the scariest-looking mathematical formula in all of mathematics (you'll see the other probable candidate in a few sections). But it's an important formula, and now, having learned a good deal of mathematical notation, we can understand what it means. It's easier if you take it piece-by-piece:

- $|A_m - L| < \varepsilon$: A_m is closer to L than the distance ε, or A_m is "close enough."
- $\forall m \geq n : |A_m - L| < \varepsilon$: In particular, it is "close enough" for any value of m greater than or equal to n. In other words, after the n-th element, all values stay "close enough."
- $\exists n : \forall m \geq n : |A_m - L| < \varepsilon$: They stay "close enough," this is, if such a value of n exists. So we only say that a sequence has a limit when we can find such a value—which is to say that A has to "get close" in the first place.

- $\forall \varepsilon > 0 : \exists n : \forall m \geq n : |A_m - L| < \varepsilon : \ldots$ It must do so, moreover, for every possible value of ε.

- $\lim_{k \to +\infty} A_k = L \leftrightarrow : \ldots$ If (and only if) the preceding is true do we say that A has a limit. The value of the limit is of course L itself.

Some other examples: the sequence (0.9, 0.99, 0.999, 0.9999, 0.99999, ...) has a limit of 1. Any number after the fourth (0.9999) is within 0.0001 of 1 itself. (See figure 5.5.) The sequence $(1, -\frac{1}{2}, \frac{1}{4}, -\frac{1}{8}, \frac{1}{16}, -\frac{1}{32}, \ldots)$ has a limit of 0, despite the fact that it swings wildly from positive to negative and back again. And the rather unenlightening sequence (4,4,4,4,...) has a limit of 4. In any of these cases, you should be able to find some number n so that every element after the nth is within, say, 0.01 of the limit.

This framework nicely captures (all right, maybe not that nicely) our intuitions about limits. More importantly, it describes a well-behaved set of infinite processes where we can use our intuitions about limits in conjunction with ordinary mathematics to get useful and meaningful results. For example, it is fairly easy to prove that if A and B are both sequences with limits (that is, if $\lim_{i \to +\infty} A_i = L_A$ and $\lim_{i \to +\infty} B_i = L_B$), then the sequence you get by adding elements of A to elements of B also has a limit, and the limit is, as one might expect, the sums of the limits ($\lim_{i \to +\infty} A_i + B_i = L_A + L_B$).

Similarly, multiplying each element of a sequence with a limit by a fixed value gives you a sequence with a new limit: the one you would get by multiplying the limit by that same fixed value ($\lim_{i \to +\infty} cA_i = cL_A$ if $\lim_{i \to +\infty} A_i = L_A$). In fact, it is possible to construct an entire algebra (in the sense elaborated in chapter 3) using limits that can then be used to calculate an unknown limit of a sequence from a group of known limits of known sequences.

There are, we should say, sequences that do not have limits. For example, there is no number that the sequence (1, 2, 3, 4, 5, 6, ...) gets and stays close to. It starts by getting closer and closer to 100, for example, but once it passes 100, it roars off into the sunset without a backwards glance, getting farther and farther away. Mathematicians will say such a sequence has no limit, that the limit "does not exist" (sometimes abbreviated d.n.e.), or that the sequence **diverges**. You will sometimes hear that the sequence "goes to infinity," but that, as we demonstrated at the beginning of this chapter, is perhaps not the most precise way to put it. "Infinity" isn't a number you can go to.

The light-bulb sequence is, crucially, another example of a sequence that diverges. Looking at the partial sums, you see that the sequence is properly stated as (1, 0, 1, 0, 1, 0, ...). It gets "close to" 1, but then zips off to become "close to" 0, and then back to 1. In knock-down argument between Patrick and Steve, Patrick was essentially claiming that the limit is 0, while Steve was claiming that it was 1. As we can now see, the right answer is that the limit isn't anything, because the sequence diverges.

5.2.3 Limits at a Point: The Epsilon-Delta Definition

So far, we've only discussed limits of sequences in which numbers get bigger and bigger and related numbers get closer and closer to something—like a dart-thrower improving with practice the more darts he or she throws. But another important concept involving limits is the idea of the limit of a function at a particular point. The idea here is that as one number gets closer and closer to something, another related number gets closer and closer to something else, even if these numbers aren't structured as an ordered sequence.

For example, if all of the four-bedroom houses within fifteen blocks of yours are worth between $100–125,000, then the appraiser will probably tell you that your four-bedroom house is worth something like that. And if all the houses within two blocks are worth from $110–115,000, then that's what yours is probably worth, too. The closer the "comparable" houses are to yours (in both location and size), the more reliable your estimate.

The same argument works for weather patterns; if you want to know what the temperature is to within ten degrees or so, any thermometer within a mile can probably tell you. If you need to know the temperature to within a single degree, you need one closer to where you actually are. If you need precision to tenths of a degree, you may need to be within inches. Weather and temperature are probably better examples of this than house prices, since you can't build a house to be smaller than the people living it in.

The mathematical framework we developed in the previous section can be easily extended to this case; instead of only one thresh-

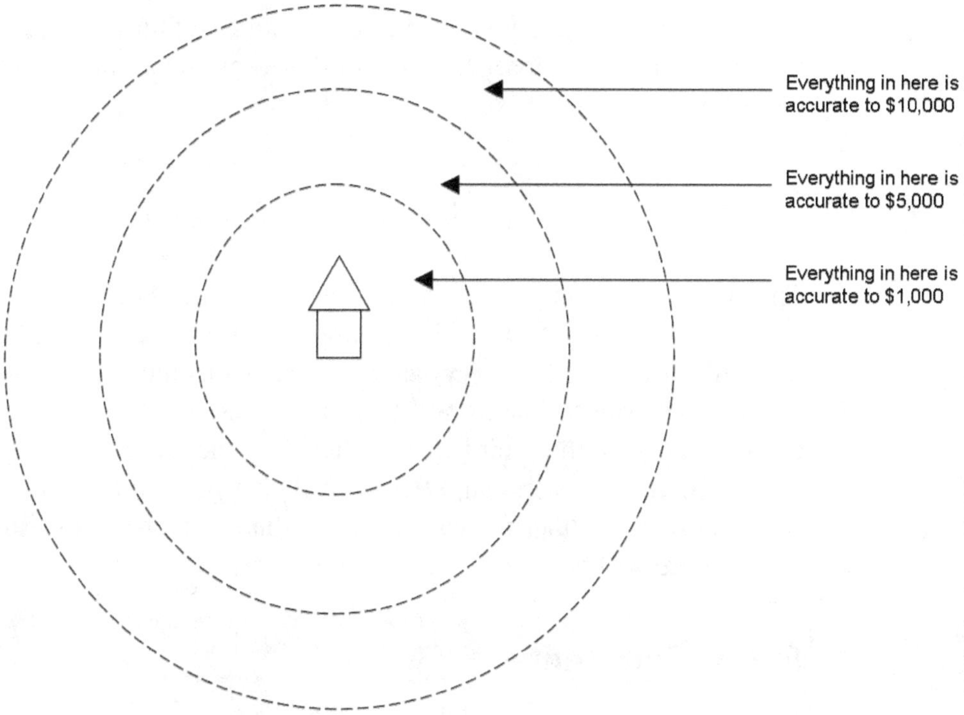

Figure 5.6: The closer you get, the more accurate you are

old (ε), we use two (the second usually uses a lower-case delta: 'δ'), and the general idea is that if you pick any number (house, location) within δ of (ie., "close to") the point of interest, the related number (price, temperature) will be within ε (again, "close enough"). Or, turning this around, if you need a temperature within ε, make sure the thermometer is closer than δ. In the formal notation, we have the following:

$$\lim_{x \to a} f(x) = L \leftrightarrow \forall \varepsilon > 0 : \exists \delta > 0 : |x - a| < \delta \rightarrow |f(x) - L| < \varepsilon \quad (5.6)$$

This is that "possibly even more scary" formula, but as with the previous one, it's not as bad if you break it into pieces. It says that the limit of the function $f(x)$ as x goes to a specific value a has a well-defined value if and only if the function gets "close enough" (closer than ε) to the limit value L when the function variable x gets "close enough" (closer than δ) to a (we'll give you a moment to read that again). If my thermometer is closer than δ, then my measured temperature will be accurate to within ε.

5.2.4 Continuity

We are now in a position to give a mathematical explanation of one of the more important concepts in applied mathematics: the idea of "continuity." We studied discrete systems in chapter 2—systems where data can be cleanly categorized into a few clear-cut categories. In a discrete system, it's either January or it's February, but it's never January-and-a-third; the defendant is either guilty or not guilty; we speak of having having zero sisters or one, but not two-fifths of a sister. By contrast, in a **continuous** system, one category blends smoothly into another with no clear line of separation (red becomes purple becomes blue, five centimeters becomes five and a half becomes six). This concept captures our intutions that a falling rock or a moving arrow moves smoothly through space instead of teleporting suddenly from point to point in a puff of smoke.

We are now in a position to state this more formally, and to relate it to the concept of a limit. In essence, we can say that a mathematical function is continuous at a particular point if and only if it has a limit at that point and that limit is the same as the function's value at that point. (A function that is continuous at all points is said to be "continuous everywhere" or simply "continuous.")

Yes, that was a bit dense. So let's unpack it. If you have a function that doesn't have a value (as when you try to divide $\frac{0}{0}$ or take $\sqrt{-1}$), it is not continuous, because it is not defined. If the function doesn't have a limit, that means that it has some sort of sudden break in it, as January turns suddenly into February instead of undergoing some kind of gradual transition. Finally, if the function has a value but the value doesn't match the limit, this again means that there's a sudden jump, a break in the function.

Most of the the functions that people work with are continuous functions simply because the world is (mostly) continuous; if you look at a succession of frames from a movie, you see that the movement is broken up cleanly into discrete images. But at the same time, you know that if you had a higher-speed camera, you would see that the spaces between the images are filled up smoothly with smaller and smaller bits. If you are at a distance x from the camera 1 second

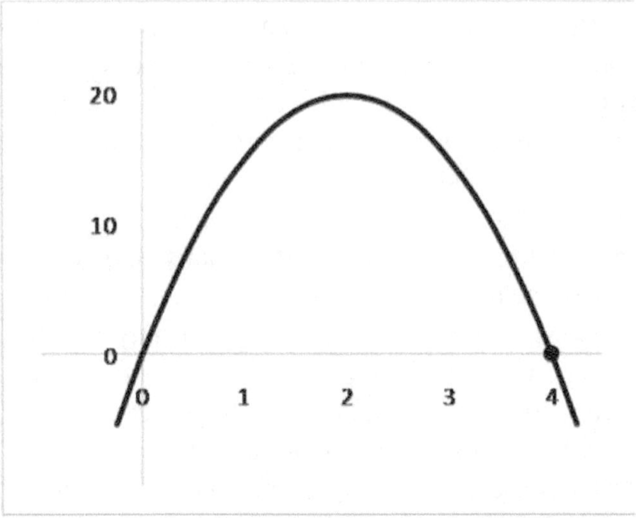

Figure 5.7: A continuous function

into the film, then you would be at a distance $x + \varepsilon$ from the camera at 1 second plus δ.

But again, not all functions are continuous. Changing the year on New Year's Eve, for example is not continuous. At 11:59, it's (let's say) 2008, heading into 2009. At 11:59:59, it's still 2008. At 11:59:59.9, it's still 2008. As we get closer and closer to midnight, it still stays 2008. Suddenly, it jumps from 2008 to 2009 (at 12:00:00 exactly). There is no time prior to 12:00 where the year is "close to" 2009, and hence the year "function" does not have a limit at 12:00.

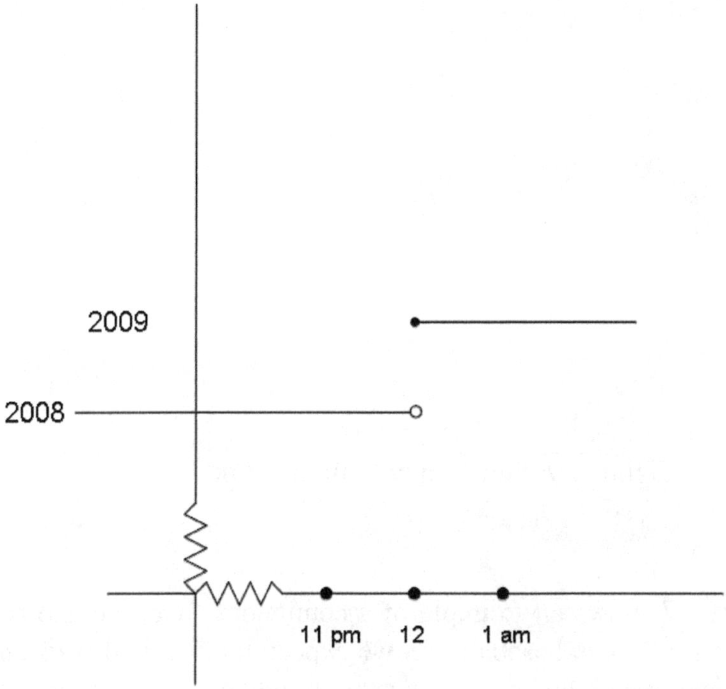

Figure 5.8: A discontinuous function

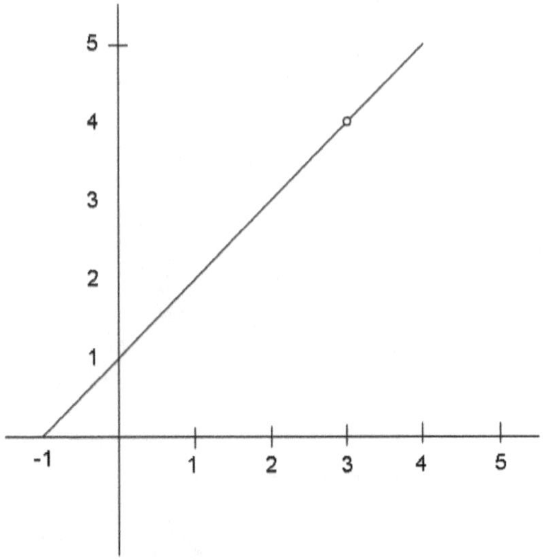

Figure 5.9: Function with limit but no value

Figure 5.7 shows an example of a continuous function: the rise and fall of a thrown baseball. As we expect, the baseball rises and falls smoothly with no breaks, gaps, or abrupt changes. By contrast, figure 5.8 shows how the year changes at midnight, with a clearly visible "gap." Finally, figure 5.9 shows a particularly interesting case in which the function has a limit but no value at the point of interest.

No value at the point of interest? Figure 5.9 is just the graph of the function

$$f(x) = \frac{(x+1)(x-3)}{(x-3)} \tag{5.7}$$

But notice that everywhere except at $x = 3$, the $(x-3)$ terms cancel out. So most of the time, we can rewrite that function as $g(x) = x+1$. If were were to let the value of x get very, very close to 3, the value of $f(x)$, as you would expect, would get very close to 4. *But at the value* 3 *itself,* something deeply odd happens. The function turns into $\frac{0}{0}$, which is mathematically illegal. Because of this, *f has no value at 3,* and so f is not continous (i.e. "**discontinous**") at 3. In essence, there is a tiny little "hole," symbolized by the open circle at the value 3—an infinitely small break in the line.

It is this bit of bizarre mathematics that enabled modern thinkers to resolve one of the longest-standing problems in physics: Zeno's paradox, or the problem of "instantaneous speed."

5.2.5 Motionless Arrows and the Definition of Speed

As we saw at the beginning of this chapter, Zeno was essentially maintaining that motion is an illusion—something inherently impossible. He also illustrated this with the idea of an arrow in flight. Considering such an arrow, we imagine time being divided into the smallest possible moments. At any such moment—at that *exact* such moment—the arrow cannot possibly be moving, since motion implies time in which to move in. The arrow is thus "instantaneously" at rest, and since time is just a collection of instants, the arrow is

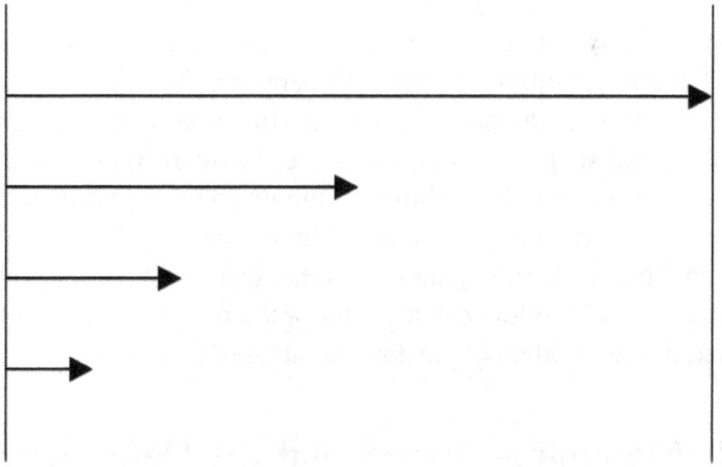

Figure 5.10: The arrow has to get halfway there before it gets there, and so on forever.

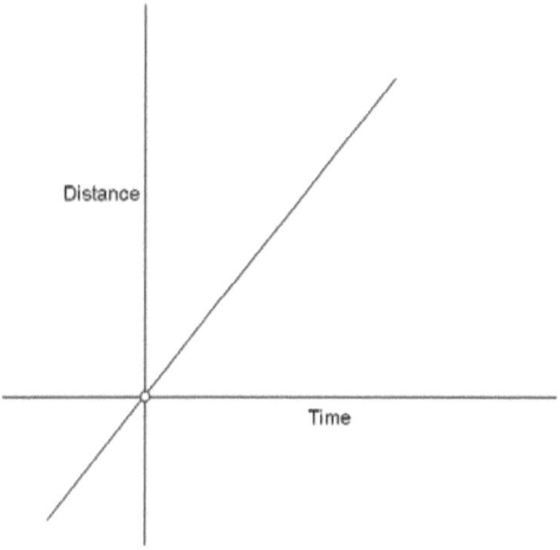

Figure 5.11: Defining "speed" as distance/time

always at rest. Hence motion does not exist. As Aristotle put it in the Physics: "[I]f everything when it occupies an equal space is at rest, and if that which is locomotion is always in a now, the flying arrow is therefore motionless" (VI:239b5).

In more modern terms, we define "speed" as distance divided by elapsed time. If I take an hour to walk 2 km, I walk at 2 km/hour. If I take 6 minutes (0.1 hours) to walk 0.2 km, I still walk at 2 km/hour. If I take 0.002 hours (about 7.2 seconds) to walk 0.004 km (4m, about the width of a typical room) . . . well, you get the idea. But if I take no time at all to cover no distance at all, how fast am I moving? After all, a snail, or a high-speed race car, could equally be said to take no time at all to cover no distance whatsoever. In each case, we

Time (s)	Distance (m)
0.00	0.00
0.50	1.25
1.00	5.00
1.50	11.25
2.00	20.00
2.50	31.25
3.00	45.00
3.50	61.25
4.00	80.00
4.50	101.25
5.00	125.00

Table 5.2: Distance covered

divide 0 by 0 to get . . . an undefined mathematical entity. This is the "hole" in the line in 5.11.

"Zeno's reasoning, however, is fallacious" (VI:9,239b5) as Aristotle correctly noted. It was not really until the nineteenth century, however, that anyone possessed the means of refuting it mathematically, since doing so requires the kind of mathematics we've been discussing. The definition of speed, like the final example in the previous section, may have a hole in it, but there is a well-defined limit value surrounding the hole.

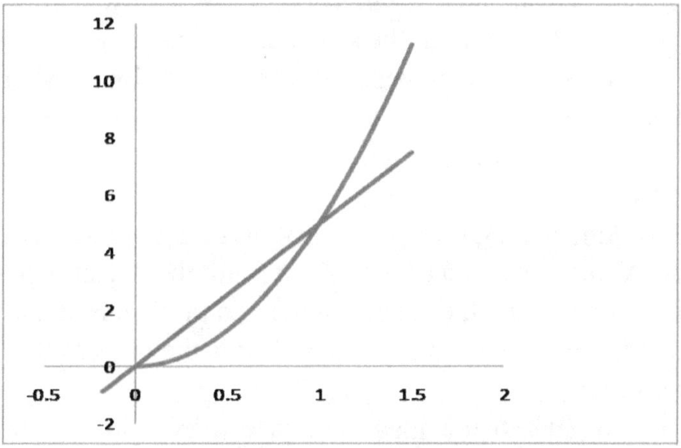

Figure 5.12: . . . and in graphical form

5.2.6 Instantaneous Speed as a Limit Process

Consider the data in table 5.2—which shows the distance of an object that is free falling—and its graphical representation in figure 5.12.[3] We can see that in the first second, the object moves from 0m to 5m—that is, a distance of 5m—in 1 second. We would therefore say that it is moving at 5m/sec during the first second. In the second second (ahem), it moves from 5m to 20m, a distance of 15m, and thus is moving at 15m/sec. (You can see these visually in the diagram: the speed of the object is simply the **slope** of the straight lines, where slope is defined as "rise over run"—the change in height divided by the change in length.) Note, however, that is is an average speed over the entire second.

These second-long averages are too coarse to give us much accuracy, however. At the exact end of the first second, the object is part of both the first second and the second second, and it's speed can't be both 5 and 15 at the same time. So how fast is the object falling *at that exact instant?*

We could try to get a better idea of its speed by using smaller intervals and getting a more accurate "average." But we can go considerably further by calculating the average speed over a short period of time—and taking *the limit of that average,* as the period becomes infinitely short. Just as we say that the polygons approximate the circle, and that the limit of the polygons *is* the circle, so we can say that the limit of the averages in this case gives us the "instantaneous speed."

[3] Yes, we are neglecting air resistance and also using a simplified value for the value of gravity — 10 m/s/s instead of the more correct 9.8.

Visually, we note that the falling object creates a curve when time is plotted against the distance fallen. This curve is, of course, a **parabola**, and we can further note that each interval creates a straight line that intersects the parabola at two points (the ends of the interval). The slope of this line we have interpreted as the change in distance divided by the change in time: i.e, the speed. As the interval becomes shorter and shorter, the endpoints become become closer and closer, until "eventually" (in the limit) they are the same point, and the line touches (and is parallel) to the curve at exactly one point. Such a line is called **tangent** to the curve, and its slope is thus the speed at that particular instant. (See figure 5.13 for an example of a tangent line.)

5.2.7 The Derivative

Questions like, "What is the speed *at that exact instant?*" seem as if they should have straightforward answers, and yet, as we hope we've demonstrated, the problem is fraught with difficult questions that relate ultimately to the notion of "the infinite." Limits make such questions tractable by giving us a mathematically well-behaved construct that we can use to perform calculations as easily as we might perform basic arithmetic operations.

In our last example, however, we went even further. Figuring out the speed *at that exact instant* turned out to be just a matter of figuring out where that tangent line lies. And indeed, finding that line is a critical matter in a truly vast set of mathemtical problems involving change. That line is so important, in fact, that it has its own name: the **derivative.**

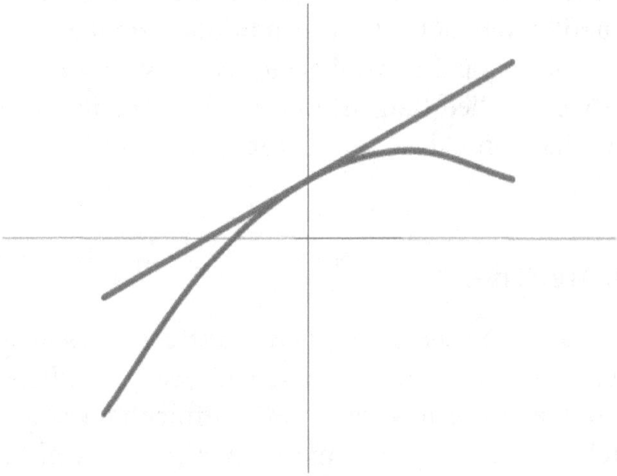

Figure 5.13: A curve and a tangent line

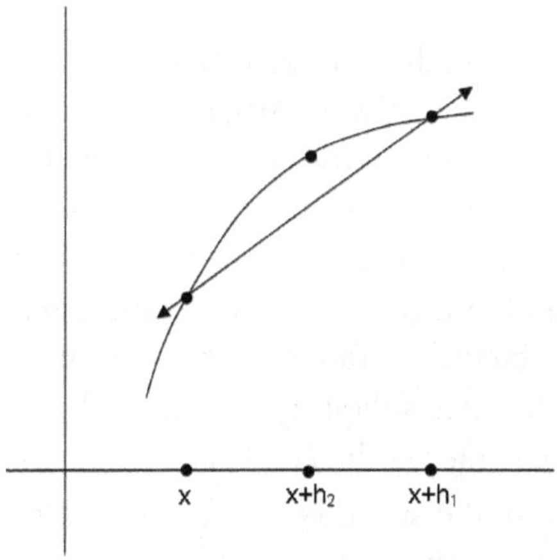

Figure 5.14: Calculating a tangent line via limits

Formally, for any function $f(x)$, the derivative of the function is defined as

$$f'(x) = \lim_{h \to 0} \frac{f(x+h) - f(x)}{h} \tag{5.8}$$

If, that is, the limit exists. Functions for which a limit does exist are called *differentiable* functions, and fortunately, most of the functions one encounters in the real world are.

It should be apparent that this is just the formalization and generalization of the falling-object example given earlier. Using the algebraic framework from earlier chapters, we let f represent the "distance fallen" function. So if x is the time of interest, $f(x)$ is the distance fallen by that time. For an interval of length h, x is the time at the start of the interval, $x+h$ is the time at the end of the interval, $f(x)$ is the distance that the object had fallen at the beginning of the interval, and $f(x+h)$ the distance fallen by the end. Therefore, the distance fallen during that interval is $f(x+h) - f(x)$ and the duration is h (as stated earlier), so the speed is $\frac{f(x+h)-f(x)}{h}$, as the formula requires. If we need to know the instantaneous speed, we want h to be as small as possible ("infinitely small" or "infinitesimal") and therefore take the limit as h goes to zero. If h were in fact zero, the value of the fraction (the speed) would be $\frac{0}{0}$, but by taking the limit we avoid division by zero. This, ultimately, is the formal answer to Zeno.

Here, there are a few different styles of notation. $f'(x)$ is often used to denote the derivative of a function, but some authors prefer to use the notation $\frac{d}{dx}f(x)$ or $\frac{df}{dx}$. Sometimes if you write an equation as something like $y = mx + b$, then y' would be the derivative of the function $f(x) = mx + b$ (you'll see this notation a lot in the next chapter.) And to be honest, we've encountered even more arcane forms of notation (particularly in older texts). It's usually obvious from context what is being discussed.

5.2.8 Calculating Derivatives

It is possible to calculate the derivative of most formulas using only the limit definition given above. In practice, there are some rules of thumb that can be used to calculate derivatives more quickly. Here are a few:

- if f is a constant value, like 3, then f' is always 0. So if a foot is always 12 inches, then we can say that the change in the length of a foot, over any period of time, is zero. To put it more plainly, if something doesn't change (i.e. is constant), then the rate of change is 0.

- if f is the product of a constant value and another function, then f' is just that constant times the derivative of the other functions. For example, if $f(x) = 12 \cdot g(x)$, then $f'(x) = 12 \cdot g'(x)$. If your houseplants' height in feet is growing at 1 foot per month, they are growing at 12 inches per month as well.

- if f is a straight line, then the rate of change is (to the everlasting relief of calculus students everywhere) the slope of the line. In this case, the rate of change is a constant and the "instantaneous" rate of change is just the rate of change. Mathematically, if $f(x)$ is just something like mx, then $f'(x)$ is just m.

- if f is the sum of two other functions, f' is the sum of the derivatives of the other functions (i.e. if $f(x) = g(x) + h(x)$, then $f'(x) = g'(x) + h'(x)$). If your legs got an inch longer and your torso an inch longer, you are a total of two inches taller.

The most important rule tells us how to handle functions of a specific format. It's a little less intuitive:

- if f is a power of a variable (like x^2 or x^{34}, or $\frac{1}{x^3}$, which is x^{-3}, or even just x itself, which is x^1), then f' is calculated by lowering the power by 1, and then multiplying by the original (unreduced) power as a constant. For example:

1. $$\frac{d}{dx}x^2 = 2 \cdot x^{2-1} = 2x^1 = 2x$$

2. $$\frac{d}{dx}x^{34} = 34 \cdot x^{34-1} = 34x^{33}$$

3. $$\frac{d}{dx}x^{-3} = -3 \cdot x^{-3-1} = -3^x - 4$$

4. $$\frac{d}{dx}x = 1 \cdot x^{1-1} = 1x^0 = 1$$

Using these rules in succession, we can calculate fairly complicated derivatives. For example,

1.

$$\frac{d}{dx}(x^3+x^2) = \frac{d}{dx}(x^3) + \frac{d}{dx}(x^2) = 3 \cdot x^{3-1} + 2 \cdot x^{2-1} = 3x^2 + 2x$$

2.

$$\frac{d}{dx}(5x^2) = 5\frac{d}{dx}(x^2) = 5 \cdot 2x = 10x$$

3.

$$\frac{d}{dx}(x^4 - x^3 + x^2 - x + 1) = 4x^3 - 3x^2 + 2x - 1 + 0$$

4.

$$\frac{d}{dx}(mx+b) = m\frac{d}{dx}x^1 + \frac{d}{dx}b = m \cdot 1 + 0 = m$$

In fact, at this point, you should be able to write down any polynomial function $f(x)$ (remember, a polynomial function is simply the sum of a whole bunch of (multiples of) different powers of a variable, like $x^3 - 2x^2 - 7x + 3$; we discussed these a bit in the algebra chapter and discuss it a bit more in the appendix) and take its derivative $\frac{d}{dx}f(x)$.

Other specific formats of functions may have other rules associated with them:

- if f is a trigonometric function (like $\sin x$ or $\cos x$), then f' is another trigonometric function. In particular:

$$\frac{d}{dx}\sin x = \cos x \tag{5.9}$$
$$\frac{d}{dx}\cos x = -\sin x \tag{5.10}$$
$$\frac{d}{dx}\tan x = \sec^2 x \tag{5.11}$$

- if f is a logarithmic function, then f' is simply $\frac{1}{x}$ divided by the **natural logarithm** of the base. If $f(x) = \log_{10} x$, then $f'(x) = \frac{1}{x \ln 10}$.[4]

- if f is an exponential function (like 2^x or 10^x) then f' is simply the f itself times the **natural logarithm** of the base. For example, if $f(x) = 2^x$, then $f'(x) = \ln 2 \cdot 2^x$. And in particular, if $f(x) = e^x$, then (since $\ln e = 1$) $f'(x) = e^x$ as well; the function is its own derivative.

For still more complicated functions, there are other rules. For example, if f is the trigonometric function called the "cotangent"

[4]Of course, this isn't very helpful if you don't remember what a "natural logarithm" is. A logarithm (the symbol for which is log) is the "opposite" of raising something to a power. So if $10^3 = 1000$, $\log_{10} 1000 = 3$ (this is usually referred to as the logarithm "to the base 10"). The "natural logarithm" is the logarithm taken to the base of the rather odd-looking number $e = 2.718281828459045235 36\ldots$. We'll defer the mystery of why this particular number is so "natural" to the next chapter.

of x [$f(x) = \cot x$], then f' is the cosecant of x, squared, multiplied by negative one [$f'(x) = -\csc^2 x$]. And yes, we looked that one up. The important thing is to realize that various shortcuts and rules of thumb exist for various classes of problems, and that it's easy enough to look up what those shortcuts are.

But to return to the main point: The important thing to note is that the derivative function exists (in most cases), and that it fundamentally represents the concept of change. For example, the numbers in figure 5.12 can be described by a simple polynomial function, $f(x) = 5x^2$ (meters). At time x seconds, the object has fallen $5x^2$ meters. At any instant, the speed of the falling object is $f'(x) = 10x$, as we have just seen. So after one second, the object has fallen five meters, and is moving at 10 m/s. After two seconds, the object has fallen 20 meters and is moving at 20 me/sec. And at zero seconds, the object has not yet fallen at all, and so is not moving (0 m/sec).

We can also take the derivative of a derivative. For example, the position of the falling object is given by $f(x) = 5x^2$, and the rate of change of the position is $f'(x) = 10x$. $f''(x)$ is therefore simply 10 ($\cdot x^0$, which is just 1 of course). Similarly, $f'''(x)$ is 0, since 10 is a constant. And, since 0 is also a constant, $f''''(x)$ is also 0.

In determining such **higher derivatives**, we are, in a sense, talking about the rate of change of a change—a matter of great interest to physicists and engineers. The derivative of position with respect to time is the speed—how fast position changes. The derivative of the derivative (more often called the **second derivative**) or the position is the rate of change of the speed, which we know as the acceleration. The **third derivative** is how fast the acceleration changes; automotive engineers know this concept as the "jerk" (a concept that doesn't show up in our example, since the "jerk" of a freely falling object is zero—the acceleration due to gravity is a constant). In this way, our original problem (the rate of change of position, or speed) can be understood as the **first derivative,** though it is rarely spoken of as such. Normally, "derivative" just means "first derivative." There's no reason in theory that we couldn't calculate derivatives beyond the third, though it is not commonly necessary.

As should be clear by now, derivatives can be used to calculate any rate of change. Economists, for instance, will often talk about the "marginal cost" of production, which is essentially the amount that costs would increase if you made one more widget. This is really just the derivative of the costs, which is another way of saying "the rate of change of costs," which is another way of saying the tangent line you get when you graph the number of units against the total costs of production.

5.2.9 Applications of the Derivative

The simplest application of the derivative is just to tell if something is going up or down. Consider the graph of the function presented in figure 5.15 (and its derivative). This is an inverted parabola—the kind you would get if you threw a rock into the air. Whenever the rock is going up, the derivative is positive; whenever the rock is going down, the derivative is negative. More generally, whenever a derivative is positive, the measured quantity is increasing.

We can similarly use the second derivative to tell us something about the shape of the curve. In figure 5.16, the curve on the left is "concave upwards," like a shaving mirror, while the curve on the

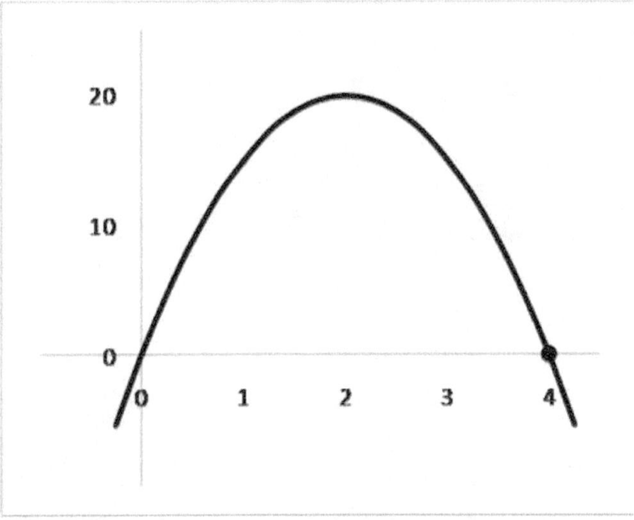

Figure 5.15: Figure 5.7 repeated

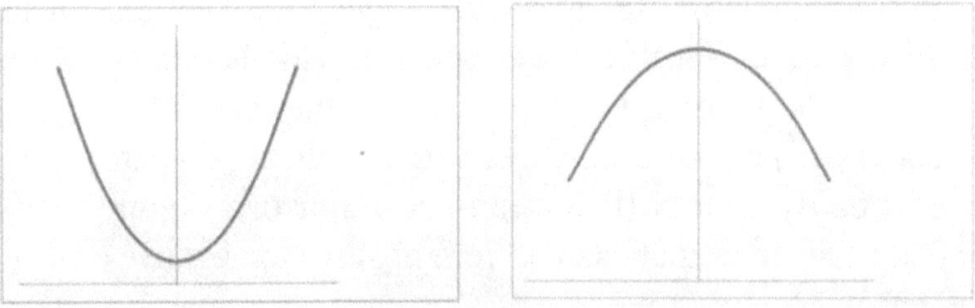

Figure 5.16: Shapes with positive and negative second derivatives

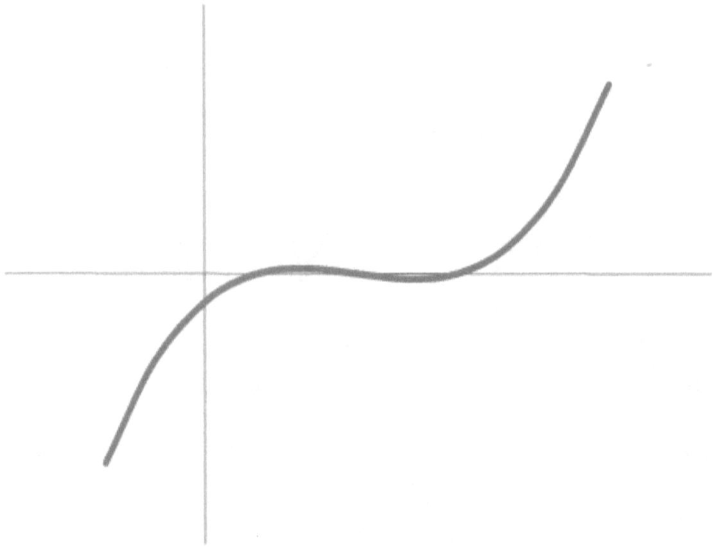

Figure 5.17: Shape with varying second derivative

right is "convex upwards" (as in a wide-angle, side-view mirror in which "objects are closer than they appear"). This is related to the sign $(+/-)$ of the second derivative; shapes that are concave upwards shapes have positive second derivatives. Shapes that are partly concave and partly convex (like a fun-house mirror or figure 5.17) have second derivatives that change from positive to negative or vice versa.

However, the points where the derivative changes sign from positive to negative (or vice versa) are the really interesting ones. First, notice that unless something really strange is going on, the only way that a continuous function (like a typical derivative) can change sign is by going through zero. (This is an important enough observation to have a name of its own: the "**mean value theorem**.") More importantly, these points are where the object changes from going up to going down (or vice versa), and thus represent the object getting to its highest point (or lowest, if we're changing from negative to positive). In ordinary terms, the function has hit the top. In technical terms, this point is called a **local maximum**.[5]

Suppose, for example, that the height of a thrown rock were given by $f(x) = 50x - 5x^2$. Notice that at 0 seconds, the rock is on the ground at height 0 (still being thrown), but at 10 seconds, the height of the rock is 0 as it hits the ground again (we'll assume we have very powerful arms, and can get 10 seconds of hang time). *How high was the rock thrown?*

Well, we know that the height of the rock achieves its maximum at the point where the derivative of height is zero. Since the height is $f(x) = 50x - 5x^2$, the derivative $f'(x) = 50 - 10x$. Do you see where we got that? Solving the equation $50 - 10x = 0$ gives $x = 5$, so the rock hits its maximum height at 5 sec.

Ah, but how high up was it? Well, at 5 seconds, the ball was $50(5) - 5(25)$ meters in the air. 125 meters. That's a very powerful arm; powerful enough to put a rock on the top of a 40 story building.

[5] Are you getting tired of all these "or vice versa"s too? If the derivative changes from positive to negative, then we have a local maximum. If it changes the other way, the function has hit bottom, or a **local minimum**. From here on out, we'll just talk about generically about maxima, and you can flip the book upside down to imagine minima.

Maybe we should try again with a more realistic arm. If the rock follows a path defined by $f(x) = 20x - 5x^2$, then its derivative is $f'(x) = 20 - 10x$, and will have a hang time of 4 seconds (still very good; that's about what an NFL punter can get), and hit its maximum at 2 seconds. This gives it a maximum height of 20 meters, which is a six-story building. So an NFL punter will typically put the ball as high in the air as a six-story building....

What's the difference? Well, the first case corresponds to a throwing velocity of 50 meters per second (calculate $f'(0)$ to see). That's about 112 miles per hour—way faster than even the best professional pitcher. The second case corresponds to a throwing velocity of 20 m/sec (or 45 mph). That's still a lot faster than either Steve or Patrick can throw, but not unusual for people who don't spend their days in libraries.

The idea of maximizing things goes beyond parabolas and thrown rocks, however. Almost anything can be maximized (or minimized), if you can frame it appropriately as a derivative problem.

For example, what's the best way to make a field? Suppose that we need a rectangular field at least 400 square meters (to grow enough of some specific plant, or perhaps to pen dragons in), and we want to fence it off. We can make a long skinny field, or a short fat one, or anything in between. But we're cheapskates and want to do it using the smallest possible amount of fencing. What's the best way?

Well, the forumula for the area of a rectangle is simply length times width. For a 400 square meter field, if the length is x, the width must be $\frac{400}{x}$ to make them multiply to the right number. And I need to fence off the top and bottom of the field (which are both fences of length x) as well as the left and right sides (which are both

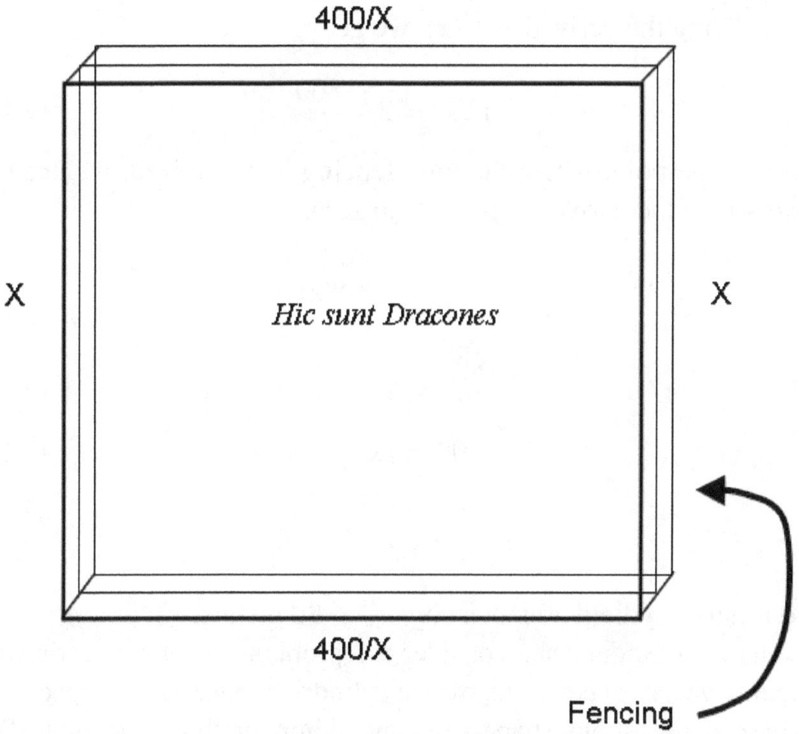

Figure 5.18: How to pen your dragon (as cheaply as possible).

$\frac{400}{x}$ meters long). I therefore need this much fencing :

$$f(x) = 2x + 2(\frac{400}{x}) \tag{5.12}$$

Calculating the derivative $f'(x)$, we get

$$f'(x) = 2 - \frac{800}{x^2} \tag{5.13}$$

To find the point where the total fencing is minimized, we need to find where the derivative is equal to zero:

$$0 = 2 - \frac{800}{x^2} \tag{5.14}$$

$$\frac{800}{x^2} = 2 \tag{5.15}$$

$$800 = 2x^2 \tag{5.16}$$

$$400 = x^2 \tag{5.17}$$

$$20 = x \tag{5.18}$$

So I make my field 20 meters by $\frac{400}{20} = 20$ meters.

Here's a harder one: consider the problem of making a tin can. Again, we can make a can, even a cylindrical one, with a capacity of 1 liter in almost any shape—tall and skinny or short and squat. But suppose I want to make the can cheaply. More specificially, what are the dimensions of a 1 liter can (which of course is 1000 cubic cm) that use the least amount of tin?

To frame this properly, we need a few formulas. The volume of the can (which is fixed, in this case at 1000) is given by the formula

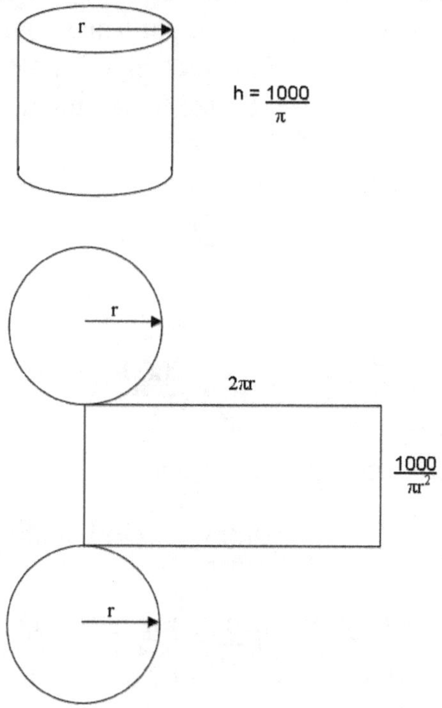

Figure 5.19: How to make a can as cheaply as possible

$1000 = \pi r^2 h$, where r is the radius of the can, and h the height. Solving for h gives us $\frac{1000}{\pi r^2}$, which we'll use in a bit.

How much metal is involved in making the can? We need the top, which is a disk of area πr^2, the bottom, which is a similarly-sized disk, and then a rectangle of height h and long enough to bend around the disk. The length of the rectangle is $2\pi r$ (the formula for the circumference of a circle). So the total amount of metal needed is given by:

$$
\begin{aligned}
f(r) &= 2\pi r^2 + 2\pi r(h) && (5.19)\\
&= 2\pi r^2 + 2\pi r \frac{1000}{\pi r^2} && (5.20)\\
&= 2\pi r^2 + \frac{2000}{r} && (5.21)
\end{aligned}
$$

Differentiating this function is easy, but tedious:

$$f'(r) = 4\pi r - \frac{2000}{r^2} = \frac{4(\pi r^3 - 500)}{r^2} \qquad (5.22)$$

A fraction can only be zero when its top is equal to zero (why?) so $f'(r)$ is zero only when $r = \sqrt[3]{\frac{500}{\pi}}$. In other words, the best size for such a can is about $\sqrt[3]{160}$ in radius (call it 10 and a half cm across) and about 10 and a half cm in height to minimize the area of the can. Of course, if you are making real cans, it's a little more complicated, as the amount of metal in the can isn't the only cost consideration. We could set up a slightly different equation, balancing the total cost of metal in the can with the cost involved in welding it together, which would be proportional to the perimeter, not the area, of the shapes involved.

5.3 Integration

Calculating derivatives (or **differentiation**) represents one of the major uses of the concept of a limit, but it has taken us a somewhat afield from our original discussion (which, you'll recall, involved the area of circle). The other major use of limits—the calculation of **integrals** or **integration**—bears more directly on the area problem. In fact, it can be used to calculate the area of highly irregular regions, as well as the more ordinary ones.

5.3.1 The Area Problem

Suppose, for example, that I have a parabola that I need to paint. Maybe I'm drawing a picture (a very tall mural) of the punter we saw in the previous section, and I want to draw attention to how high the punt is by painting everything under it in bright yellow.[6] The question, of course, is how much area is under that parabola?

We can approach this problem using the method of exhaustion described earlier. We could, for example, paint it in rectangular candy-stripes, each 1m wide. Since we know that the painting is 4 m wide, we will need four stripes. The first will start at 0m and end at 1m; the second will start at 1m and end at 2m, and so on. We also know the

[6] We're humanists, not artists.

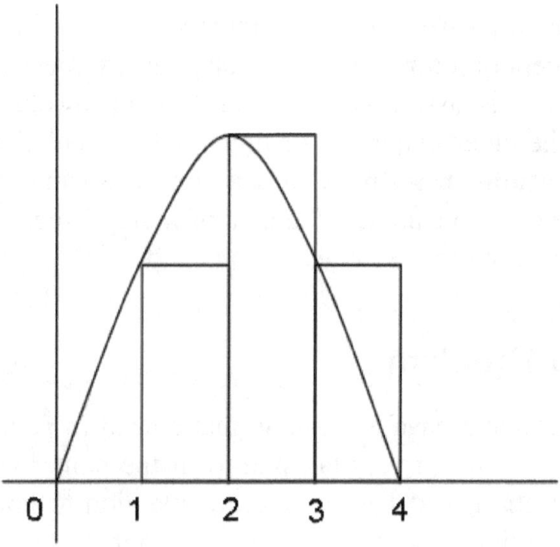

Figure 5.20: Painting a mural

formula for the area of each rectangular stripe: base times height. But how tall will each stripe need to be?

They'll have to be different heights, since we're painting under a curve. The function f of the height of the parabola will answer that one just fine. If $f(x) = 20x - 5x^2$, then the first stripe, the one starting at 0, will be 0m tall. The second stripe will be 15m tall. The third stripe, starting at 2, will be 20m tall, and the third, starting at 3, will be 15m tall. (See figure 5.20). The total area, then, would be the sum of the four areas, each of which have base of 1 and varying heights as shown, so: $1m \cdot 0m + 1m \cdot 15m + 1m \cdot 20m + 1m \cdot 15m$, or $50m^2$.

Of course, those candy stripes don't cover all the area under the curve. But if we drew narrower stripes, say, only 50cm (0.5m) across, we could fit more stripes under the curve and measure the areas more exactly. (Figure 5.21) If we drew stripes only 10cm across, or 1cm, we would be more accurate yet. This is, as we noted, simply the "method of exhaustion" performed with rectangles instead of with regular polygons. But it should be clear where we are going with this. As with our previous examples, if we let the number of stripes grow infinitely large (or alternatively, let our stripes become infinitely thin), then the area of the stripes (which we know and can calculate) will tend as a limit to the area underneath the curve.

5.3.2 Integrals

Informally, the "definite integral" (or more loosely, the **integral**) of a function $f(x)$ is simply the limit of the sum of such rectangles' areas. More formally, the integral of f from a to b, written

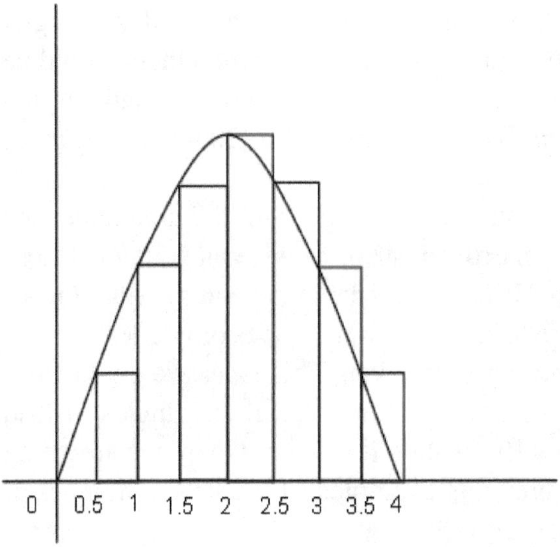

Figure 5.21: Painting a mural in thinner stripes

$$\int_a^b f(x)\,dx$$

is the area of the region bounded on the left by the x-axis, the lines $x = a$ and $x = b$ and f itself.

This unusual S-shaped notation denotes the limit of areas of the infinitely many infinitesimally small rectangles that together comprise this area. So if we know how to calculate the area of a rectangle, we can calculate the area of almost any shape imaginable.

The actual calculations for this area would be extremely tedious. Fortunately for us, one of the finest mathematicians who ever lived was able to determine that we don't need to do these calculations at all. The relationship that demonstrates why has been called the most important mathematical idea of the last two thousand years.

5.4 Fundamental Theorem of Calculus

Historically, calculus has been divided into two major branches. Differential calculus concerns itself with the calculation of derivatives (which were historically called "differentials") and the calculation of the slopes of tangent lines. Integral calculus (which was at one point thought to be part of the mostly unrelated problem of areas) concerns itself with the calculation of integrals. For centuries (more or less from the Greeks to Isaac Newton and Gottfried Liebniz), these two problems were approached in isolation from one other.

It was Newton's teacher (Isaac Barrow, 1630–1677) who showed that these two problems are in fact the same problem. Specifically, differentiation and integration are inverses of each other, in the same way that multiplication and division, or addition and subtraction, are inverses of each other. This insight was formalized by Newton and Liebniz into one of the crowning intellectual achievements of the modern era: the **Fundamental Theorem of Calculus**.

Despite the extraordinary amount of lateral thinking necessary for noticing this, the result can be expressed very simply. If f is a function, and g is the integral of f, then g', the derivative of g, is equal to f. Or alternatively, if g is a function, and f is the derivative of g, then the integral of f (written $\int f$) is g.[7]

What this means in practice is that any method of calculating derivatives can be used "in reverse" to calculate integrals. For example, the little shortcuts we presented in section 5.2.8, like the fact that the derivative of a sum is the sum of the derivatives, can be reversed to capture the integral of a sum as the sum of two integrals (or vice versa, if necessary). We can intuitively describe the speed as the rate of change of position, or we can (alternatively) describe a position as the sum of all the previous speed changes applied over time. The Fundamental Theorem of Calculus validates both of those views and captures, formally, the relationship between the two, essentially

[7]The formal statement of the Fundamental Theorem of Calculus is a little bit more complex than we have suggested. We omit detailed discussion of it here in order to stay true to our plan for a conceptual introduction, but the full explanation is, of course, widely available for those interested. Similarly, the actual details and methods for calculating integrals, especially integrals of tricky functions, are beyond what we intend to cover here—and in any event, can often be looked up in books like Gradshteyn and Ryzhik's *Table of Integrals, Series, and Products* if you need them.

creating "calculus" as a single, unified subfield of mathematics.

So, remember the mural (figure 5.21)? What *is* the area we need to paint? Well, if you remember, the formula for the curve itself was $f(x) = 20x - 5x^2$, and we calculated the derivative $f'(x)$ as $20 - 10x$. But we want to find a function $g(x)$ such that $g'(x) = f(x)$. We can use the trick mentioned for differentiating polynomials (multiply by the exponent and reduce the exponent by one) in reverse to find such a function. The derivative of $10x^2$ is $20x$ (do you see why?). The derivative of $\frac{5}{3}x^3$ is $5x^2$. So the derivative of $g(x) = 10x^2 - \frac{5}{3}x^3$ is $f(x) = 20x - 5x^2$.

To calculate the area under the curve, we just evaluate $g(x)$ at the edges of the drawing, which is to say, at points 0 and 4. The value of $g(0)$ is 0, and the value of $g(4)$ is $\frac{160}{3}$. Subtracting one from the other yields a total area of $\frac{160}{3}$ or a little over 53 m^2. So our original estimate of 50 square meters wasn't bad at all.

5.5 Multidimensional Scaling

Calculus is key to almost any analysis that involves the words "minimum," "maximum," or even "best fit." You've already seen how a cost-conscious engineer can apply calculus to the problem of making a tin can as cheaply as possible. In this section, we'll talk about a more abstract application, a very powerful visualization and analysis technique, called **multidimensional scaling** (or MDS).

At its heart, multidimensional scaling is a method of creating a "map" based on measured or perceived distances, such that things that are similar are close together and things that are distinct are far apart. This provides and easy way to visualize the variation among items and how items cluster, without necessarily having a pre-defined visualization space.

The mathematical basis is fairly straightforward. First, the user is assumed to have, for every pair of objects, a measure of the the "distance" between them. In some cases, this could be an actual distance, or it could be a measure of similarity based on judgement—for example, the number of people in a focus group study who put two objects in the same pile. If you have n objects, this will of course give you an n by n matrix as we saw in an earlier chapter.

We know from geometry that any three points can be put into an exact relationship in a plane that preserves distances perfectly. Just as two points define a line, three points define a plane, and in general, n points define an $n-1$ dimensional space in which these distances can be preserved. Needless to say, humans have a bit of trouble visualizing seven-dimensional space. What we want to do is find a two-dimensional space that preserves distances as best we can, so that we can see the relationships between and among the various points.

And the key to doing this involves calculus. MDS creates a configuration of points in a two-dimensional space and assigns each object a single point in this space. As points in this new space, they are separated by a (genuine) distance on the page. The trick is to minimize the "stress:" the overall difference between the genuine distance on the page and the similarity-based distance given by the user. If points X and Y are far apart on the page but close in the matrix (or vice versa), we have a problem. More formally, if we define δ_{ij} as the distance listed in the matrix between objects i and j, and

d_{ij} as the ordinary distance between the points that represent i and j, then we want to minimize the total across all pairs of $(\delta_{ij} - d_{ij})^2$. The word "minimize" is what makes this a calculus problem. What we're after is a point where the derivative is zero.

Unfortunately, it's a rather a difficult calculus problem, since it is often not possible to solve for this point directly. But the same insights that let us see how calculus works let us see how we can use computers to solve this problem via iterative approximation (which is just a fancy way of saying "let's take a wild guess and then improve it over and over again"). If you can figure out, for example, which point has the highest stress associated with it and move it slightly in the proper direction, that change in direction will improve the overall stress. We can then do this until that point is no longer the one with the highest stress, and do the same thing with whatever point has the (next) highest stress. Eventually, we'll get to a point where no small changes to anything make any difference to the overall stress, which is a pretty good example of what "where the derivative is zero" means. Of course, this is laborious and time-consuming—not to say boring—work to do by hand, but computers are great at doing this kind of thing. In fact, computers can use even more complicated, but faster ways to do MDS.

The power of MDS can be seen from table 5.3 and figure 5.22.[8] Table 5.3 lists the distances between some cities. Given such matrix, can we reconstruct where they are in relation to each other?

The answer is : yes, and no. See Figure 5.22.

[8]Based on data from *http://www.stat.psu.edu/ chiaro/BioinfoII/mds_sph.pdf.*

	Boston	NY	DC	Miami	Chicago	Seattle	SF	LA	Denver
Boston	0	206	429	1504	963	2976	3095	2979	1949
NY	206	0	233	1308	802	2815	2934	2786	1771
DC	429	233	0	1075	671	2684	2799	2631	1616
Miami	1504	1308	1075	0	1329	3273	3053	2687	2037
Chicago	963	802	671	1329	0	2013	2142	2054	996
Seattle	2976	2815	2684	3273	2013	0	808	1131	1307
SF	3095	2934	2799	3053	2142	808	0	379	1235
LA	2979	2786	2631	2687	2054	1131	379	0	1059
Denver	1949	1771	1616	2037	996	1307	1235	1059	0

Table 5.3: Distances (in miles) between major US cities

The one thing that MDS doesn't know about is orientation. So the computer did a very good job of reconstructing the relationships between the cities, but managed to get the north-south axis flipped.

So why would a humanist care about MDS? Well, one possibility is to use it in conjunction with GIS systems to locate things that we're not sure about, such as lost historical sites. If we could construct a matrix of travel times (based on contemporary reports) that described how long it took to get from a known place (such as the Tower of London) to a place we have since lost track of (Castle Anthrax?), we could construct a map like the one above and then superimpose it on a real map of England to get a better idea of where the castle stood.[9]

[9]We can use measurements to create abstract maps that visualizise other useful aspects of the data, and both authors use the technique frequently when dealing with text data. Patrick has worked extensively in authorship attribution, using computational techniques to measure the "distance" between authors. MDS allows him to cluster and visualize authors and how they group based on stylistic similarity, which may or may not match similarities in genre or time. But this same type of analysis would allow historians to look at the volume

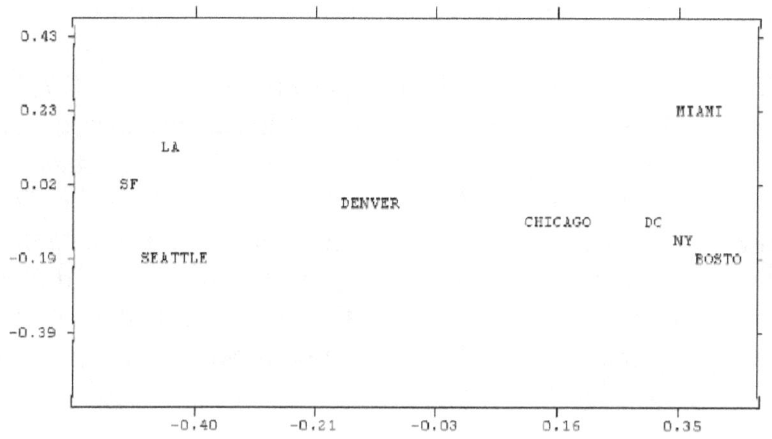

Figure 5.22: Map of US cities from multidimensional scaling

5.6 Calculus Made Backwards

At the beginning of this chapter, we noted that we were going to try to get at the concepts of calculus "through the back door." What we meant by that will perhaps be clearer when we note that the original proof of the Fundamental Theorem of Calculus did not avail itself of

of shipping between any two ports, and use this volume as a measure of how "close" they are economically. Would it be surprising to learn that colonial New York has closer economic ties to Liverpool than to colonial Savannah? And this could in fact be true, even if almost no ships sailed directly between the two, as long as (for instance) there was a lot of traffic from New York to London and from London to Liverpool ....

the idea of limits, and in fact, could not have done so. Newton and Leibniz did their work in the seventeenth century—long before the theory of limits was discovered by Augustin-Louis Cauchy (1789–1857) in 1821. Much of modern instruction in calculus directly reverses this pattern. In a typical university curriculum today, students are usually given a course in "pre-calculus" that discusses the idea of limits (and associated concepts such as continuity), and then this course continues with calculus proper (in which the ideas of differential calculus and the Fundamental Theorem of Calculus are set forth). Integral calculus, and the calculations of volumes of geometric forms, is usually deferred until a second or sometimes even a third course.

Historically, though, the question of area was the first to be studied (by the ancient Greeks, as we mentioned). After that, the question of tangent lines and derivatives was taken up by Archimedes, and then extended (much further) by Indian and Arabic mathematicians, who also created the idea of a derivative, in the first millennium. The Fundamental Theorem was proven by Newton and Liebniz, as we mentioned, but it would be another fifty years before the formal foundations were finally established by Cauchy. This may at least partially explain the frequent lament (made by teachers as well as students) that calculus books often read like they were written backwards. In some sense, they really were.

Early in this chapter, we described calculus as "the mathematics of change." We still think that's an apt description, and one that renders its applications in science and engineering as obvious as they are manifold. That the humanities are themselves concerned with change suggests, perhaps, a point of intersection with those fields that attempt to join humanistic inquiry with computational methods, and indeed, advanced methods such as data mining, machine learning, and statistical natural language processing frequently require methods drawn from calculus. But we cannot resist noting the ways in which calculus itself is a fit object of humanistic concern—as much an achievement of the modern world as its grandest works of art. Some say, despite the occasional strong candidate, that the great explanation of calculus—one as approachable as it is astounding—has yet to be written. We eagerly await another book (written, we suspect, either by a humanist-turned-mathematician or its inverse) that places calculus properly within the cultural context of modernity. In fact, we suspect that the former won't succeed until the latter performs its necessary work.

6 Differential Equations

In the previous chapter, we discussed the mathematics of change as expressed in the calculus, and called it one of the most brilliant ideas in mathematical history. And this is good news, because the analysis of change—including both the causes and the effects of change—is one of the key intellectual components of scientific inquiry as such. Humanisitic inquiry, too, is mostly about trying to deal with multiple forms of change and sorting out what's going on in an extremely complicated system called "human culture."

But most real-world systems, whether cultural or "natural," possess a level of complexity that is truly formidable. For the scientist, understanding complex systems is less about finding the trajectory of some object through space, and more about understanding the underlying factors that govern highly dynamical systems like climates, animal ecologies, and quantum phenomena, in which thousands (or millions) of elusive factors and underlying variables are at play. For the humanist, nearly every problem takes this essential form. What factors conspired to end monarchical government (or human slavery, or epic poetry, or photo-realistic painting) in so much of the modern world? What factors led to the rise of the English novel? What factors led to the Reformation?

The study of such systems (and the differential equations that stand at their core) is a well-established part of scientific inquiry, but similar investigation of "humane systems" is extremely rare. Indeed, the idea that the kind of change we see in cultural systems can be modeled in a mathematically tractable manner surprises many people. Economics (a field which we still consider a humanities discipline) represents one way, but even the economist acknowledges that the humanistic problems outlined above weren't merely a matter of money circulating through a society. In an age of "big data"—including big cultural data—applying the theory of dynamical systems (via differential equations) to data drawn from cultural instead of physical artifacts might well revolutionize our understanding of the human condition.

If we sound a bit vague on this point, it is perhaps because this kind of investigation—quantitative analysis of cultural phenomena—is a truly new field. And unfortunately, we must, at the risk of quoting Barbie, point out that the math is hard. We'll not lie to you; this is a challenging chapter. It's the sort of material that few people outside of advanced mathematics and the hard sciences ever try to come to grips with. And of all the chapters in this book, this one has the least obvious relation to the kinds of mathematics that humanists are inclined to want to work with right now. We present this chapter in the hope that it will be the kind of mathematics that humanists will want to work with in the future.

From a strictly mathematical point of view, our task is to discover a way to identify factors that underly changes in complex systems—factors that will often include the current state of the system itself.

6.1 Definitions

To get a sense of why this is hard, set a cup of coffee and a cold beer out on the counter and watch how the temperatures change. The beer will get warm while the coffee cools. But to model this mathematically is surprisingly tricky, because while it true that hot things will cool, and hotter things will cool more rapidly, in mathematical terms, this means that the change-in-temperature (which, you'll recall from chapter 5, is the derivative of temperature) *is related to the temperature itself.* This makes determining the temperature at any given time and how it changes a quite challenging task. At the same time, it illuminates many other important processes in the physical world, and perhaps the cultural world as well.

The hotter the coffee, the faster it cools. This situation creates what mathematicians call as a **differential equation**: an equation that describes a relationship between an unknown function and one or more of its derivatives. In this case, the physical setup can be described by Newton's Law of Cooling: the rate of change in the temperature of an object is directly proportional to the difference in temperature between the object and its surroundings. We can formalize this by the equation we are about to present. In this equation, x represents the (current) temperature of the coffee, while x' or $\frac{dx}{dt}$ represents the instantaneous rate of temperature change. We also need to know the temperature of the room (which we will call x_r and assume remains constant), and we will use the letter k to represent a currently unknown constant.

Before we get to k, though, let's make sure we understand what "directly proportional" means. Basically, it means that two functions are "in proportion" no matter what scaling factor you use. If you are twice as tall as your favorite nephew when measured in inches, you will also be twice as tall when measured in centimeters, meters, or miles, because the difference is just a constant multiplier. If you buy twice as much gasoline, you will pay twice as much, because the cost per gallon is a constant that applies proportionately to every gallon equally. In mathematical terms, x and y are directly proportional if and only if there is some k (the price per gallon) that relates x and y as:

$$x = k \cdot y$$

Notice, though, that it doesn't really matter what k is; if the price of gasoline goes up, it will have a different price per gallon, but it will still have a unique and proportional price per gallon. By contrast, temperature in Fahrenheit is not directly proportional to temperature in Celsius, since you can't apply a constant multiplier to get from 0 to 32 degrees.

And what is x? We've already said that it is "the temperature of the coffee." But since the coffee is cooling, x can't be a simple number. For x to change, x must be a function, and specifically a function of time. We could say that the temperature "now" is $x(0)$, while the temperature one hour from now will be $x(1)$. Our task, then, is to describe the function that satisfies our constraints (in this case, Newton's Law of Cooling).

Newton's Law of Cooling can be expressed as:

$$\frac{dx}{dt} = -k \cdot (x(t) - x_r) \qquad (6.1)$$

or alternatively

$$x'(t) = -k \cdot (x(t) - x_r) \tag{6.2}$$

So k (which we assume is positive) represents the rate of heat transfer (which would serve to illuminate the fact that a cup of coffee in an insulated mug will not cool as fast as a cup of coffee in a simple steel cup). The smaller k is, the slower the temperature will change; if k were 0, then it would fulfill the dream of mathematicians everywhere by providing perfect insulation that would keep coffee warm forever.[1]

If $x(t) > x_r$, then the temperature difference between the coffee and the room (at time t) is positive, so the change in temperature is negative (reflecting the fact that hot objects cool). If $x < x_r$, then the change in temperature is positive, and you can watch your cold beer get warmer. If $x = x_r$, then the temperature difference is zero, so the change in temperature is also zero—an object at room temperature will stay at room temperature. So our intuitions are captured by this framework.

Population growth provides another useful example of a differential equation. If you assume that a certain percentage of the population is pregnant at any given instant (for some animals like rabbits,

[1]Notice that if $k = 0$, $x'(t)$ will also $= 0$. But $x'(t)$ is just the change in temperature, so the temperature does not change (which would have alien archeologists digging through the rubble of our failed civilization to find a still-warm cup of coffee). In more mathematical terms, if $x'(t) = 0$, the only functions whose derivatives are 0 are constant functions. So the temperature of the coffee is constant always and forever, which would undoubtedly astound the alien archeologists even more.

this seems to be as high as 100%), then the change in population (P) is that percentage multiplied by the population itself, or

$$\frac{dP}{dt} = k \cdot P \tag{6.3}$$

where k is another constant of proportionality, taking into account the prevalence of pregnancy, the size of litters, and the survival rate. If you know that population changes, you don't need to be reminded that P is a function of time and not a simple number.

But things can get much more complicated rather quickly. What if we are trying to understand population growth in a population that includes predator/prey relationships. As we have just seen, left to its own devices, the population of rabbits will grow in accord with equation 6.3 (a disturbing bit of mathematics for areas experiencing the effect of invasive species without natural predators). A population of foxes in isolation will decrease as the foxes starve to death. But in an environment where both foxes and rabbits exist, every time a fox encounters a rabbit (which will happen as a product of the number of foxes and of rabbits), the population of foxes will increase slightly and the population of rabbits will decrease slightly. Since we don't know by exactly how much, we'll just use a variable k to represent the exact value until we get the biologists' reports.

How often will such encounters between a fox and a rabbit occur? Well, the more foxes there are, the more often that will happen, and the more rabbits there are, the more often that will happen. A simple idea is to treat these as happening at some rate proportional to the number of rabbits multiplied by the number of foxes. We can capture this with the following family of differential equations, where P_f and

P_r represent the populations of foxes and rabbits, respectively:

$$\frac{dP_r}{dt} = k_1 P_r - k_2 (P_r \cdot P_f) \tag{6.4}$$

$$\frac{dP_f}{dt} = -k_3 P_f + k_4 (P_r \cdot P_f) \tag{6.5}$$

But how do we actually solve these problems? How will the temperature of the coffee really behave? If it is at 100C at 8am, and at 30C at 9, was it warmer or colder than 65C, the halfway point, at 8:30? Although we hinted at the solution in the previous chapter, we'll now pursue it in more detail and give some guidelines for solving differential equations in general.

6.2 Types of Differential Equations

As we have seen with calculus, one of the primary mental difficulties with differential equations is simply the number of types of problems out there. Calculus students are typically called upon to memorize many different formulas and methods: one for polynomial functions, one for trigonometric functions, one for exponential functions, one for composite functions, and in more rarefied fields of mathematical endeavor, some considerably more obscure varieties. The same thing is regrettably true for different types of differential equations. Depending upon the form of a differential equation, people need to use different methods to solve them, and some forms don't have nice clean solutions at all, irrespective of the method one uses. So in order to understand differential equations, we first need to engage in a bit of taxonomy and terminology.

6.2.1 Order of a Differential Equation

In the previous chapter, we discussed higher-order derivatives, and the concept is similar here. The **order** of a differential equation is simply the highest derivative involved in the equation. All three examples in section 6.1, for example, involved only first derivatives and are therefore first-order equations, but the equation

$$y'' = -10y \tag{6.6}$$

(which might have come from a study of a spring or a swinging pendulum) is a second-order differential equation. Remember that the notation y'' means the second derivative of y, and is equivalent to the expression $\frac{d^2y}{dt^2}$.

The equation

$$y'''' = y + 16y'(y'') - \sin(y''') \tag{6.7}$$

is a fourth-order equation and quite possibly too difficult to solve by hand.

As you might expect, the smaller the order of a differential equation, the easier it is to solve (all else being equal). But as with calculus proper, low-order equations (first and second degree) are fairly common.

6.2.2 Linear vs. Non-Linear Equations

We can also classify differential equations as being **linear** or **non-linear**. In a linear differential equation, no two derivatives are multiplied or otherwise mangled in a way that would violate proportionality. Most of the examples above are linear differential equations, with the notable exception of equation 6.7. Equation 6.7 is non-linear for two reasons: first, the first derivative y' is multiplied by the second derivative y'', and second, the third derivative y''' is mangled by the non-linear sine function.

Linear equations are a lot easier to solve than non-linear ones (again, all else being equal); the theory of linear differential equations is also a lot better developed than the theory of non-linear differential equations. Fortunately, linear equations are common enough to be worth solving. For this introduction, we'll stick to linear equations.

6.2.3 Existence and Uniqueness

In light of these different categories (many of which seem to require completely different solution methods), the single most important question to ask is whether or not all the different (applicable) methods are likely to get the same answer. Mathematicians usually phrase these in terms of **existence** and **uniqueness** (a distinction by no means limited to differential equations). In essence, we're dealing with two closely-related questions: does a solution exist, and is that solution unique?

Obviously, if no solution exists at all to a problem, it is pointless to try to look for one; confirming that a solution exists before embarking on a lengthy search is therefore only common sense. But more significantly, if there are many possible solutions, two different methods might find two different solutions—and miss each other's solutions, in a sort of mathematical "blind men and the elephant" situation. On the other hand, if the solution is unique, then any two methods that find *any* solution will automatically both find *the* solution (i.e., the same one). Existence and uniqueness proofs are thus part of the bread-and-butter of working mathematicians, but particularly so with differential equations.

Fortunately, most of the existence and uniqueness theorems we will need have already been proven for us, and unless you are really into the deep mysteries of mathematical formalism, you can just take their word for it. But if you decide to take the easy way out, there is still a catch. To understand exactly what "existence" and "uniqueness" mean in this context, we first need to figure out exactly what we're trying to solve and what a "solution" looks like.

6.2.4 Some Preliminary Formalities

As we said before, in all the examples we've discussed, the "variable" (whether temperature, rabbit population, or something else) is not just a simple variable, but a function that describes how a particular aspect of reality varies over time. If P_r is "the population of rabbits," it's a time-dependent parameter: $P_r(0)$ is how many rabbits there are right now, while $P_r(-1)$ is how many there were a little while ago (which, depending on the units, could be last month, last year, or a minute ago), and $P_r(1)$ is how many there will be at some point in the future. In general, then, $P_r(t)$ is a time-dependent function, and $P_r'(t)$ or $\frac{dP_r(t)}{dt}$ or $\frac{d}{dt}P_r(t)$ are equivalent ways to describe the (instantaneous) rate of population change.

Our solution to describing the rabbit population (or any differential equation) will likewise be a function. And once we have that function, we can evaluate it at any time or location we want in order to get the exact population right now (or last year).

We will therefore adopt the following notational conventions when speaking of these abstract functions. The function $y(t)$, which we will usually just abbreviate as y, is a function that depends only upon the variable t (which will usually represent time in our examples). The symbols y', y'', y''', and so on (or alternatively $\frac{dy}{dt}, \frac{d^2y}{dt^2}, \frac{d^3y}{dt^3}, \ldots$) take their standard meanings of the first, second, third, and higher derivatives of y with respect to time. So when we say that we want to solve the differential equation $y' = 0$, that means we are to find a function $y(t)$ such that $\frac{dy}{dt} = 0$.

Now thinking back to section 5.2.8: Do we know any functions whose derivative is 0? In fact, we do: "If f is a constant value, like 3, then f' is always 0." So if we let $y(t)$ be a constant function (like 3; i.e. $y(t) = 3$), then $y' = 0$ as required. But this is likewise true for *any* constant value, not just 3. The functions $y(t) = 61$, $y(t) = 0$, and $y(t) = -14$ all satisfy our differential equation as well. This means that our solution defines not a single function, but a family of functions. We can describe this family by saying that "$y(t) = C$ for any constant value C." This sort of "family-resemblance" is common in differential equations as well. Because many different functions will have the same derivatives (differentiation is a many-to-one function), many different, but related, functions will usually solve a given equation. We can, however, usually show that these functions are all related mathematically and give a way of choosing a particular function from the family. In this case, we can say that the function y is "unique up to a constant," meaning that if we pick a specific value for C (or have it picked for us by our data), our value for y is truly unique.

The usual way we find this sort of specific value is by using a **boundary condition**. For example, if we know the temperature $y(t) = C$ (our forever-unchanging cup of coffee), then we can just look at what the temperature is *now* to know what the temperature will be *then*. Usually, the problem statement will specify one or more points (typically at the beginning of the system, hence the "boundary") exactly and the equations can be used to predict how things change away from the boundary. If the ever-warm cup of coffee is 30C right now, and the temperature never changes, then it will also be 30C tomorrow. Equation 6.46 gives a more realistic example where the cup actually cools, and we have to use the boundary conditions to determine how much.

6.3 First-Order Linear Differential Equations

The simplest type of differential equation to solve is the first-order linear differential equation. Technically speaking, any function of the form

$$y' + a(t)y = b(t) \tag{6.8}$$

where a and b depend on t, but not on y, is such an equation. There's a rule that lets the experts simply look at the values of a and b and write down the solution—but we'll have to warm up to it.

6.3.1 Solutions via Black Magic

First, consider the differential equation

$$y' = y - 1 \tag{6.9}$$

which we can rewrite as

$$y' - y = -1 \tag{6.10}$$

(Note that in this case, $a(t) = b(t) = -1$; so this is indeed a first-order linear differential equation, and a rather simple one.)

Despite its simplicity, solving this equation quickly gets us into some serious math. Let's start by picking the low-hanging fruit: remember that the derivative of any constant function $y(t) = C$ is 0. So if we let $y(t)$ be an appropriately chosen ("guessed") constant, we might get lucky. After consulting the spirits and reading the entrails of some animals, Steve guesses that the function $y(t) = 1$ is a solution. Did he get it right? It appears so. y' in this case would be 0, since y is a constant function. Substituting in the above equation, we get

$$0 = y' = y - 1 = 1 - 1 = 0 \tag{6.11}$$

Unfortunately, many people's knowledge of math, and of differential equations, stops at this level (if it rises this high in the first place); the Ph.D-equipped soothsayer (or computer) is simply asked to produce the answer, which we hope is correct and complete. If the soothsayer is good enough, this may promote accuracy, but it does not promote understanding. So once you feel confident of your ability at least to recognize a solution and to check the soothsayer's work, let's try to develop a general solution to this type of problem. (In this specific case and with the usual authorial omniscience, we can go further and state that Steve did in fact miss some possible solutions—we'll see which in section 6.3.2.)

6.3.2 A Simple Example

Possibly the simplest example of a differential equation that arises in practice is the population formula:

$$y' = ky \tag{6.12}$$

Of course, this isn't an equation by itself, but an equation schema. A person actually studying population change would have an actual value for k that describes the particular situation she's looking at (bacteria reproduce faster than rabbits, which reproduce faster than humans, which reproduce faster than elephants). For simplicity's sake, let's assume that her (empirically-derived) value of k is exactly 1. In that case, the equation becomes

$$y' = y \tag{6.13}$$

or alternatively

$$\frac{dy}{dt} = y \tag{6.14}$$

(the change in $y(t)$ is exactly $y(t)$ itself.) This is just like asking, "what function has itself as its own derivative?" We have already hinted at the answer in section 5.2.8 (the exponential function e^t). But how do we get to that answer?

We use a technique called **separation of variables** (that mostly involves algebra mixed with a little calculus). By manipulating the equation to get all the terms involving y on one side, and all the terms involving t on the other, we can get to the point where we can take integrals of both sides:

$$\frac{dy}{dt} = y \tag{6.15}$$

$$\frac{1}{y}\frac{dy}{dt} = 1 \tag{6.16}$$

$$\frac{1}{y}dy = 1dt \tag{6.17}$$

This is notation that, strictly speaking, we have not seen before; the "dy" and "dt" are not really variables as much as placeholders representing the "infinitesimal" change in both time (t) and population (y). About the only thing we can do at this point is integrate both sides, which requires calculating $\int 1\, dt$ (which is just t),[2] and

[2]More accurately, it is t plus a constant C_1, since any function of the form $t + C_1$

calculating $\int \frac{1}{y}dy$. Remember from the previous chapter that the derivative of the logarithm function $\ln y$ is $\frac{1}{y}$. so the integral again will be $\ln y + C_2$. We thus have the following:

$$\int \frac{1}{y}dy = \int 1\,dt \tag{6.18}$$

$$\ln y + C_2 = t + C_1 \tag{6.19}$$

$$e^{\ln y + C_2} = e^{t + C_1} \tag{6.20}$$

Since $a^x \ldots a^y = a^{(x+y)}$, we can replace the added exponents with a multiplication.

$$e^{\ln y} \cdot e^{C_2} = e^t \cdot e^{C_1} \tag{6.21}$$

Now, raising e to the natural logarithm of any number simply returns that number itself (since raising a number to a power and taking the logarithm are inverse functions). Hence $e^{\ln y}$ is just y.

$$y \cdot e^{C_2} = e^t \cdot e^{C_1} \tag{6.22}$$

$$y = e^t \cdot \frac{e^{C_1}}{e^{C_2}} \tag{6.23}$$

We also notice that the expression $\frac{e^{C_1}}{e^{C_2}}$ is pretty meaningless, since C_1 and C_2 (as we said before) can be anything. In fact, this expression really just means "some number that could be anything." In

---

will have a derivative of 1. C_1 can really be anything we like at this juncture, a point to which we will return in a bit.

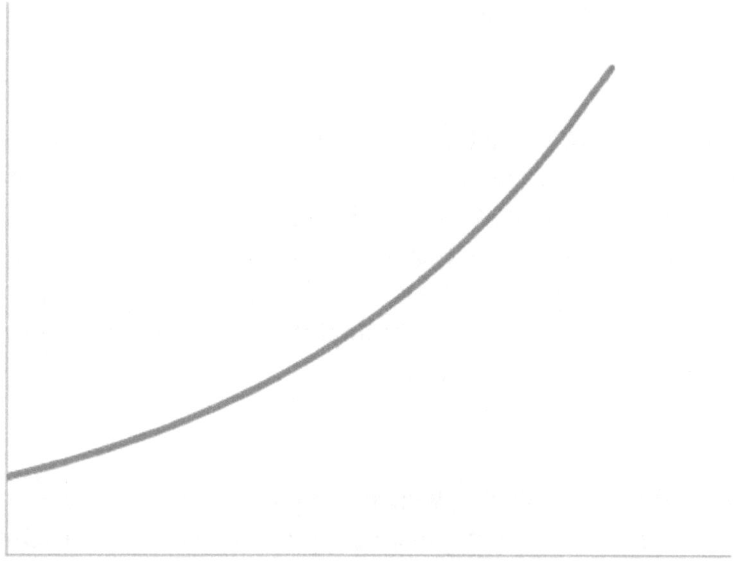

Figure 6.1: Population growth curve

this light, we have the following family of solutions to the differential equation:

$$y = e^t \cdot k \tag{6.24}$$

which basically all look like figure 6.1.

Any exponential of the form e^t, or any multiple of that form (like $3e^t$ or $75e^t$ or even $-2e^t$) is thus a potential solution to our population problem. Population, or any system described by equation 6.12, will change according to that function.

Backing away from the math for a moment, this simply says that "if the rate of change of the population is proportional to the population, then the population growth will be an exponential curve." An "exponential curve" is just the term for the shape of figure **??**. More important, thought, is the other half of the statement. We're assuming here, in our model, that the only important factor in population growth is the current population size.

So what's the value of k? Well, this again is known to the researcher. At time $t = 0$, the population would have to be $y(0)$. But $y(0) = e^0 \cdot k = 1 \cdot k = k$, so k is simply the starting population. In general, solutions to differential equations will be "unique" only up to some parameter k—but only one point in the family will satisfy a *specific* starting constraint in the real world. Our general solution is "unique up to a constant," and we can figure out the value of k by looking at our data.

Let's take a related problem—one where the constant of proportionality isn't simply 1. Suppose our researcher found that the constant were instead a negative value, $-\lambda$ (which of course is a variable which we can presume represents an actual number). The same argument applies:

$$\frac{dy}{dt} = -\lambda y \tag{6.25}$$

$$\frac{1}{y}\frac{dy}{dt} = -\lambda \tag{6.26}$$

$$\frac{1}{y}dy = -\lambda\ dt \tag{6.27}$$

$$\int \frac{1}{y}dy = \int -\lambda\ dt \tag{6.28}$$

$$\ln y + C_2 = -\lambda t + C_1 \tag{6.29}$$

$$e^{\ln y + C_2} = e^{-\lambda t + C_1} \tag{6.30}$$

$$e^{\ln y} \cdot e^{C_2} = e^{-\lambda t} \cdot e^{C_1} \tag{6.31}$$

$$y \cdot e^{C_2} = e^{-\lambda t} \cdot e^{C_1} \tag{6.32}$$

$$y = e^{-\lambda t} \cdot \frac{e^{C_1}}{e^{C_2}} \tag{6.33}$$

$$y = e^{-\lambda t} \cdot k \tag{6.34}$$

This, again, yields a family of curves, as shown in figure 6.2. Any member of this family is a potential solution.

We can determine k by looking at the starting population. If we know that the value at $t = 0$ is, say, 47, then we simply ask which of the family of curves goes through the point (0,47). So the solution is still the same shape: an exponential. And, in fact, most reasonable first-order linear differential equations have this shape, although the base of the exponential might vary (note that $e^{-\lambda t} = (e^{-\lambda})^t$. So if

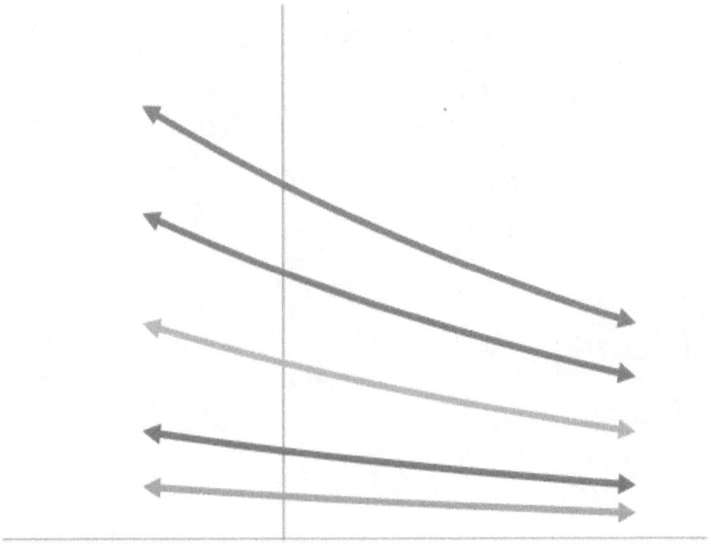

Figure 6.2: A family of falling exponentials (negative growth)

$a = e^{-\lambda}$, our final equation is equivalent to $y = k(a^t)$ (with the base a instead of e).

Does this mean that our soothsayer was wrong when he told us that the solution to $y' = y - 1$ is $y = 1$? We can certainly suspect so. Solutions to differential equations seem to come in families with expressions like e^k, but our soothsayer gave us only one member of the family: the member where the value of k is 0 and the exponential disappears. But that may not be—in fact, probably isn't—the member we had in mind.

The examples in this section have all been of **homogenous** equations, equations where, going back to the original definition (equation 6.8), $b(t) = 0$. The solution for non-homogenous equations is a little more tiresome, but still works out to be related to the exponential[3] With this general answer in hand, you can see how a mathematical expert can simply write down the family of solutions for any first-order linear differential equation, and apply the starting conditions to determine the specific values of the relevant constant.

[3]The actual formula for the solution to non-homogenous equations is $y = e^{-I} \int e^I b(t)dt + ke^{-I}$, where $I = \int a(t)dt$. Note that if $b(t) = 0$, the first half of this equation disappears (thus illustrating why homogenous equations are easier). It also puts us in a position to deliver a final smackdown to our soothsayer from section 6.3.1. He correctly identified the first part of this formula, but not the second. In general, $1 + ke^t$ is a solution for any value of k, including 0.

6.4 Applications

The applications of first-order differential equations are many and varied. Basically, any time you have a description of how something changes, either in terms of the number of something that are already out there, or in terms of something else unrelated like sunspot activity, you can generate a differential equation to describe it. In simple terms, if how much you get depends on how much you already have, a first-order differential equation can be used to model it.

Let's quickly move through a series of examples in which such equations occur. In each case, the thing to notice is the way in which one pattern or description of change is being related to another.

6.4.1 Naive Population Growth

Have we done this one to excess yet? Typically, the number of new members of a population are some fraction of the current population—more potential parents equates to more actual parents equates to more children. This can be simply described by equation 6.35, as in our example.[4]

$$P' = rP \tag{6.35}$$

An interesting variation occurs if you have some constant influx of people from some other source (through immigration, perhaps).

[4]This is sometimes called the Malthusian model of population growth, after Thomas Malthus (1766–1834), who first formalized it. Note that it is essentially the same as equation 6.12.

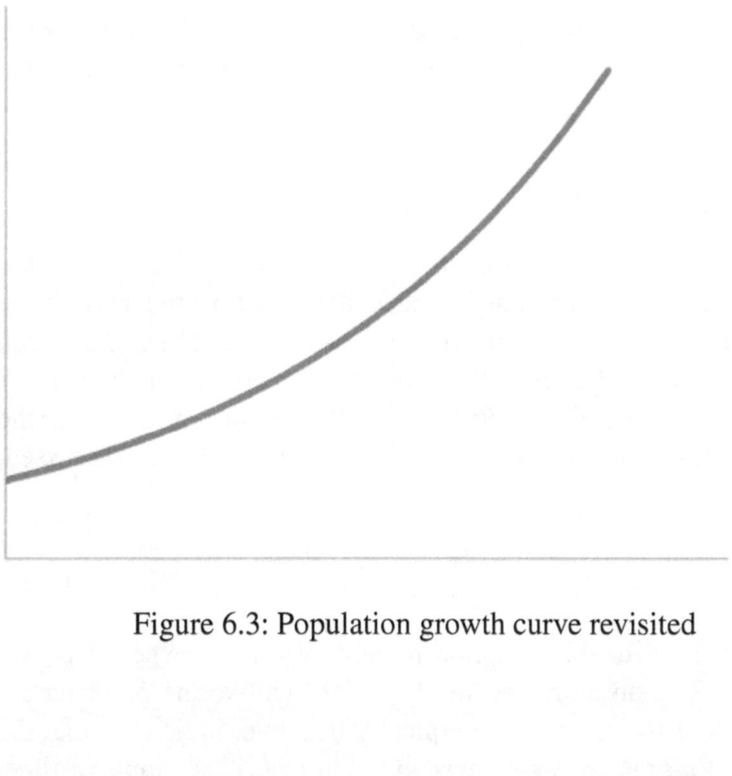

Figure 6.3: Population growth curve revisited

For example, if g is the population growth factor, but we also get f immigrants coming over the borders at every time interval, then we can modify equation 6.35 to

$$\frac{dP}{dt} = gP + f \tag{6.36}$$

This is a non-homogenous equation, of course, so it will no longer be the pure exponential of the previous section. It is still amenable to solution.

6.4.2 Radioactive Decay

Radioactive decay is another simple example of a differential equation at work; modern physics predicts that a radioactive atom has a fixed and constant chance of decaying over any equal-sized interval. In this situation, of course, the change (decrease) in the number of atoms will be directly proportional to the number of atoms in the sample. This yields the simple equation we have already discussed, where

$$\frac{dy}{dt} = -\lambda y \tag{6.37}$$

The solution to this equation remains $y = e^{-\lambda t}$, where t is still time, and λ is an atom-specific constant. Unlike the previous examples, here the exponent is typically negative, indicating that the number of atoms decreases over time. In fact, in $\frac{\lambda}{\ln 2}$ units of time, the number of atoms will be reduced by 50%—thus yielding the well-known concept of a radioactive "half life."

6.4.3 Investment Planning

If you continuously put D dollars per year into a savings account that pays interest at rate r, how much will you have at 65 (or whenever you decide to retire)? As before, interest is paid on the principle balance at a fixed proportion (the interest rate) of the balance, so the change in principle is partly proportional to the balance itself. The steady stream of deposits, however, creates another part of the change—one that is constant and does not vary with the account balance. Differential equations again:

$$\frac{dP}{dt} = rP + D \tag{6.38}$$

Sharp-eyed readers will note the similarity between this equation and equation 6.36; in fact, these two equations are identical in form except for the change of variables. Solving this equation (and using the appropriate starting condition $P(0)$ for the initial balance in the account) allows financial planners to estimate how much a given investment scheme will be worth at various points in the future.

Of course, this is something of an oversimplification of the real world; most if not all mathematical models are. In this case, we can pinpoint the oversimplification in the assumption that the interest rate is a constant r that will not change over time. In the actual world, interest rates can fluctuate wildly. Indeed, banks can collapse, in which case the our entire principle simply vanishes. But we can still use this model several times to attempt to determine a range of values; if we expect the return on an investment to be between 6–9% over the long term (and that the bank will not collapse over that time), then using the formula with $r = 0.09$ will give us an estimate of the maximum value, while $r = 0.06$ will give us the minimum. Any investment planner will be happy to solve this differential equation for you if they think you will be investing with them. In fact, that is precisely what they are doing; their business model depends upon it.

A related problem, of course, is that of annuity pricing. Here, I wish to buy a fixed investment (and pay $P(0)$ for it), with the intention of making fixed withdrawals from it over the next twenty or so years. This is more-or-less the same equation, except that instead of adding D deposits, I will subtract W withdrawals:

$$\frac{dP}{dt} = rP - W \tag{6.39}$$

As you can see in figure 6.4, the balance of the account follows a falling exponential, starting at some value (the price you pay today) and ending at zero twenty years from now (at which point the account will be closed and you hope to no longer need the money). Again, depending upon the amount of withdrawal you wish to make, and the interest rate you can obtain, your investment advisor will tell you how large a check you need to write today to guarantee yourself a steady retirement income over the next twenty years.[5]

[5] Actually, that's only one possible path the balance can take. If the initial balance is high enough, then the interest rP will not be offset by the withdrawals that I make, in which case the account balance will continue to climb. Wait a bit for this to become an important digression.

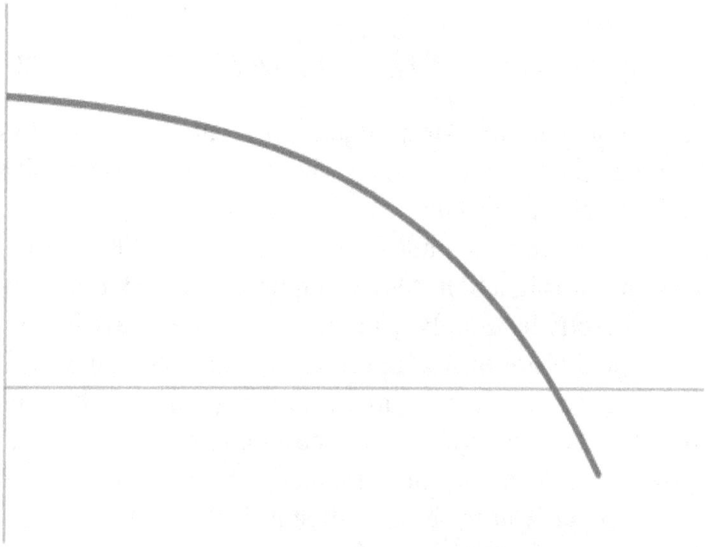

Figure 6.4: Withdrawing money from an account or annuity

6.4.4 Temperature Change

Newton's Law of Cooling has already been discussed; hot objects get cooler and cool objects get hotter in direct proportion to the difference in temperature between the object and the ambient temperature A (or earlier, t_r) of the surroundings:

$$\frac{dy}{dt} = -k(y-A) = -ky + kA \tag{6.40}$$

This non-homogenous first-order linear differential equation has the usual solution, and depending upon what we know, we can solve it for any of the relevant constants with enough data.

Remember our coffee problem? If our coffee was 100C at 8am, and 30C at 9, what was it at 8:30? We can now answer that question. We could do it directly by simply using the formula we gave earlier, but it's simpler (and more fun) to use a trick. Since the rate of heat exchange is related to the temperature difference (and not just to the temperature), we can pretend that our thermometer is broken and always registers 20C low. So the (assumed) room temperature of 20c would measure at 0, and the hot coffee at 100C would measure at 80. We thus get the modified equation

$$y' = -ky \tag{6.41}$$

to which we already know the answer: $y = e^{-kt} \cdot c$.

But we haven't solved the problem yet. We don't know the value of k (how well insulated the cup is), and we don't know the multiplicative constant c. But our problem gives us the information we need.

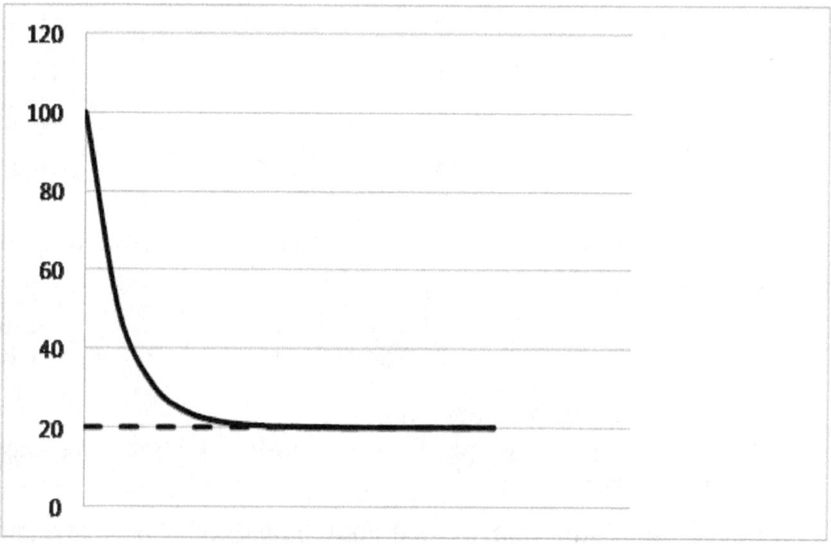

Figure 6.5: Change in coffee temperature

In particular, we know that after 0 hours (at 8am), the coffee is at 80 degrees (on our broken thermometer). Hence $y(0) = 80$. But $y(0) = e^{0k}c = e^0 c = 1c = c$, so $c = 80$. We also know that after 1 hour, the coffee is at 10C, so $y(1) = 10$. But $y(1) = e^{-1k} \cdot 80$. Thus

$$10 = 80e^{-k} \tag{6.42}$$
$$10e^k = 80 \tag{6.43}$$
$$e^k = 8 \tag{6.44}$$
$$k = \ln 8 \sim 2.07944154 \tag{6.45}$$

Knowing k (from the boundary conditions), we can rewrite equation 6.41 as

$$y' = -2.07844154 \cdot y \tag{6.46}$$

with $y(0) = 80$ as our boundary condition.

We want to know what $y(\frac{1}{2})$ is. From our work above, we know that it's $e^{-2.07944154 \cdot 0.5} \cdot 80$. The value of $e^{-1.04}$ is about .35 (we don't need quite precision yielded above), and $.35 \cdot 80 = 28$. Thus our broken thermometer would read 28 at 8:30, and the actual temperature would be 48. This is substantially cooler than 65, which is halfway between 100 and 30. In fact, it looks like the coffee did 65% of its first hour's cooling in the first half hour.

One interesting application of this law and this type of analysis is in medical forensics (where we're trying to estimate the time of death). A normal human body has a temperature of about 37C (98.6F). After death, it will typically cool (over a matter of a few hours) to the ambient temperature, 20C or so indoors, and almost anything outside. Suppose that a coroner is called to the scene and measures the temperature of a dead body at 30C at midnight, and then re-measures the temperature at 25C two hours later (assuming the ambient temperature is 20C throughout.) At this point, one can use the two values $y(0) = 30$ and $y(2) = 25$ to solve the equation and determine the value of k, how fast this specific body loses heat in this specific circumstance. But once he's done that, he can now use this equation to predict at what time the body temperature was 37C (the time will naturally be negative, and in this case will be around -1.5). This means, of course, that the body was alive an hour-and-a-half before midnight, so we conclude that the time of death was at about 10:30pm.

6.4.5 More Sophisticated Population Modeling

The straight-up Malthusian model of population growth is not often used (except when populations are tiny). Indeed, Malthus himself is probably best known for his failed predictions of food riots, wars, pestilences, and starving masses standing on each other's shoulders as the world population increases without limit. The reason this doesn't happen is because the population can't grow without limit. In practical terms, one must consider not just the birth rate, but also the death rate (and how the death rate changes with changes in population).

First, we note that as the population grows, there will be competition for resources. Similarly, some members, even if they do not die, will be unable to successfully reproduce. Thus the death rate will increase and the birth rate will decrease as the population increases. If we say that the death rate is $d + aP$ and the birth rate is $b - cP$, the rate of total change in population becomes:

$$
\begin{gathered}
(b - cP) - (d + aP) \\
b - cP - d - aP \\
b - d - cP - aP \\
(b - d) - (a + c)P
\end{gathered}
$$

Thus the revised differential equation to model the population should be

$$P' = [(b - d) - (a + c)P] \cdot P \tag{6.47}$$

More conventionally, this formula will be rewritten. We will consider $b - d$, the "natural" birth rate without population effects, minus the natural death rate, to be r_0, the "natural" reproduction rate. We then define K to be $r_0/(a + c)$. Using this, we can rewrite the previous equation as

$$P' = r_0(1 - \frac{P}{K})P. \tag{6.48}$$

or

$$P' = \frac{r_0}{K}(K - P)(P) \tag{6.49}$$

This equation is sometimes called the **logistic equation** and for

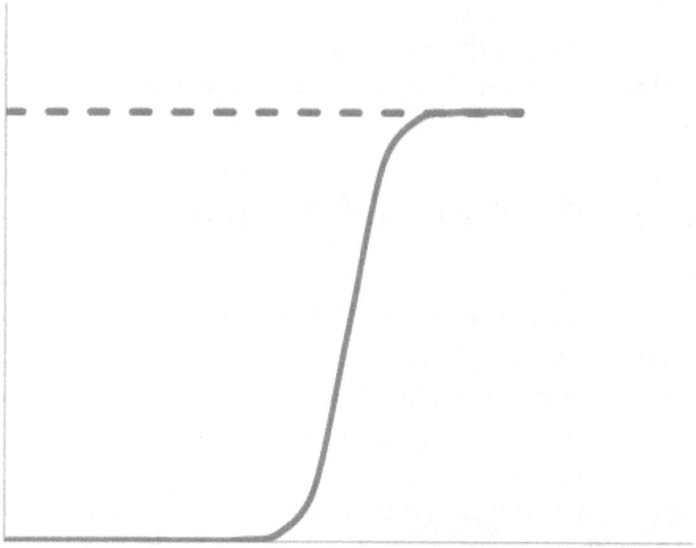

Figure 6.6: Population curve with finite capacity (logistic curve)

that reason, this model is called the logistic model.[6] With a population near 0, $\frac{P}{K}$ is close to 0, and the reproduction rate is near r_0, but as P is close to K, the reproduction rate becomes almost zero. In fact, if $P > K$, the reproduction rate becomes negative, as more organisms starve than reproduce.

The general shape of the curve generated by this is rather S-shaped,

[6]If you compare this with the Malthusian model (equation 6.35), you see that r is a constant in his model, but in the logistic model, r $(= r_0(1 - \frac{P}{K}))$ varies with the population as well as the natural carrying capacity of the environment K.

starting close to 0 and rising slowly, then rising more quickly as the population grows (nearly exponentially), then slowing down again as the population nears K and the death rate overtakes the birth rate. (See figure 6.6).

6.4.6 Mixing Problems and Pollution

Consider the following problem: we have a 1000 liter tank filled with pure water, from which we extract 200 liters of water per minute for some industrial process. To keep the tank full, we have another pipe that delivers 200 liters per minute into the tank, but (unknown to us), the water coming in is polluted and has 1 g/liter of unobtainium. How much unobtainium is in the tank?

Again, we can set this up as a differential equation to tell us how the concentration changes over time. Perhaps obviously, $y(0) = 0$ at the point where we started polluting, and it will get higher over time. But what can we use for a process description?

The unobtainium has to go somewhere. The change in the amount of unobtainium in the tank is equal to the amount flowing in, minus the amount flowing out. We know the amount flowing in: it's simply 1 g/liter times 200l or 200g (per minute). The amount flowing out is a little trickier, since it depends on the amount already in the tank. If we make the (possibly incorrect) assumption that the tank is perfectly mixed, then the rate at which it leaves the tank is the flow rate times the concentration in the tank.

In other words, if there are y grams in the tank, then the concentration is $\frac{y}{1000}$ g/l. With a flow rate of 200l/min, this means that $\frac{y}{1000} \cdot 200$ g/minute is leaving the tank (which we can reduce to $\frac{y}{5}$).

Thus, the total change is:

$$y' = 200 - \frac{y}{5} \tag{6.50}$$

This formalization can be used to solve many related problems; for example, a factory dumping mercury into a lake will raise the concentration in the lake, and we can determine from the concentration and the dumping rate just how long it has been happening and how much longer until the mercury levels become dangerous.

6.4.7 Electrical Equipment

Most people are familiar with batteries and resistors; batteries, of course, supply power to an electrical circuit, and resistors consume power, converting it into heat (which is essentially how your toaster works). Specifically, there is a relationship between the voltage across a resistor and the current (I) going through the resistor given by Ohm's law: $V = IR$.

Not as many people are familiar with capacitors and inductors. Capacitors are typically two small plates separated by a gap which stores electric charge, and which respond not directly to voltage, but to changes in voltage. Inductors are typically small coils of wire that respond to changes in current by generating a magnetic field. In particular, the current flowing through a capacitor is the derivative of the voltage, and the voltage across an inductor is the derivative of the current. This means, among other things, that both capacitors and inductors can store energy. But what happens when you get a circuit that contains both resistors and capacitors/inductors?

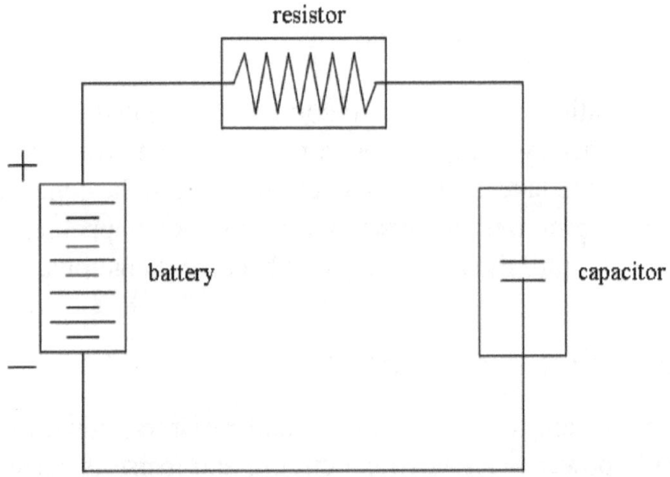

Figure 6.7: A simple resistor-capacitor circuit

The answer, although a bit out of our scope, is that you get a differential equation. Electrical engineers are typically taught in their first year how to analyze RC circuits (circuits with resistors and capacitors) and LR circuits (circuits with resistors and inductors) as first-order differential equations. For example, the simple RC circuit in figure 6.7 is described by

$$R\frac{dI}{dt} + \frac{1}{C}I = \frac{dV}{dt} \tag{6.51}$$

where R, I, and V are resistance, current, and voltage, and C is the

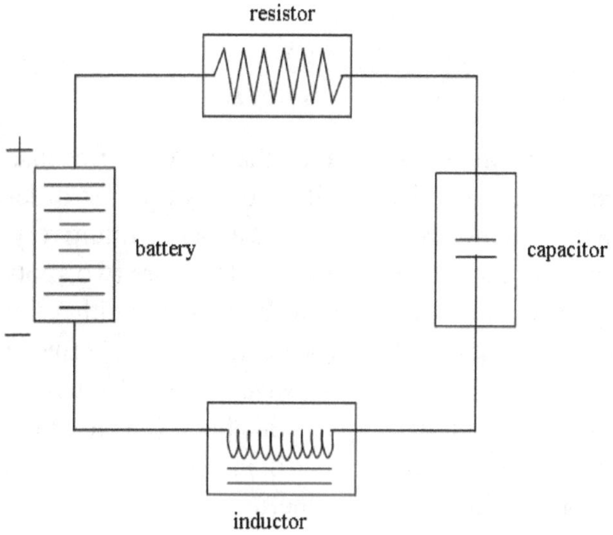

Figure 6.8: A simple resistor-capacitor-inducter circuit

capacitance of the capacitor (essentially a measure of how fast it can respond to electrical changes). In practical terms, the energy of the battery will go into the capacitor at a gradually decreasing rate; if the battery were to die, the capacitor could continue powering the circuit for a (typically small) amount of time.

But what about RLC circuits (figure 6.8) (circuits with both a capacitor and an inductor, with L representing "inductance"—the speed at which the inductor can respond to changes in current). Well, in this case, the equation is more complicated:

$$L\frac{d^2I}{dt^2} + R\frac{dI}{dt} + \frac{1}{C}I = \frac{dV}{dt} \tag{6.52}$$

where $\frac{d^2I}{dt^2}$ denotes the **second derivative** of the current. Because this equation involves a second derivative, it is no longer a first-order differential equation and can no longer be solved as such. Intuitively, the behavior is much more complicated, because there are two spots that can store energy. Energy can go from the battery into either the capacitor or the inductor, but it can also wash back and forth between the capacitor and the inductor in a kind of wave. A full discussion of this point is probably beyond the scope of this book (Patrick took a two semester course sequence on this stuff in college), but this should at least give you the flavor of the issue.

6.5 Sethi Advertising Model

Of course, most of the readers of this book are expected to be humanities scholars, not electrical engineers—it's understandable if your interest in "capacitors" is limited. But the same techniques that describe how resistors respond to current can also describe how foxes respond to rabbits and vice versa. Or, for that matter, how humans respond to advertising. Given a reasonable set of assumptions, is it possible to model how a company's market share varies over time with their advertising budget? The most reasonable form for such a model is to look at market share, and to model the change in market share as a function of other aspects.

According to the model developed by Suresh P. Sethi, there are three important aspects to advertising. The first is simply the response of the public, especially the unsold public, to the advertisement, a factor that will tend to increase market size. The second is the natural tendency of people to forget about ads (or to respond to the ads of competitors instead), a factor that decreases market size. The third factor is essentially a random effect, one that can go either way and covers unforeseen and unforeseeable effects.

The actual Sethi advertising model (simplified a bit)looks like this:

$$X' = rU\sqrt{1-X} - \delta X + \sigma \tag{6.53}$$

where X is the current market share (at time t), and X' is therefore the change in market share. The unsold public is of course $1-X$, the amount of people who don't already buy your product. U represents the rate of advertising, essentially how much money or TV time you're buying, while r is the effectiveness of the advertising (selling hearing aids on the radio or advertising literacy classes in the *NewYorker* may not be a particularly effective tactic). δ is a "decay coefficient" describing the natural tendency of market share to fall off, and σ is just a random factor.

The first term, $rU\sqrt{1-X}$, shows, initially, that the more you spend on advertising and the more effective your advertising is, the more your market share increases. At the time time, though, the greater your market share is initially, the less you will be able to increase it — if you already have 95% of the custom, the number of additional people who *can* buy your stuff is only 5%. (We saw a similar term to this in the sophisticated population growth example; the underlying argument is exactly the same. Just as an ecosystem can't have more people than its carrying capacity, similarly you can't have more than 100% market share.)

The second term, δX, shows that the more customers you have, the more you stand to lose to competitors. The third term of course just says that "stuff happens" and circumstances beyond your control may give you a better or worse market share.

We're not going to solve this equation here. It would be boring and cumbersome. But we wanted to show it to you. From an economic point of view, this formula is nice because it provides a mathematically tractable way to predict the effectiveness of an advertising campaign. In general, most companies don't want to have a 100% market share, or more accurately, aren't willing to spend the huge amounts of money it would take to get that large a market share. As you can see from the formulation above, to get your market share above 99% ($X > 0.99$) would require a tremendous amount of investment because the multiplier $\sqrt{1-X}$ would be so small, requiring a huge U to offset it. Intuitively, we understand that that last stubborn customer might need a huge amount of convincing. But this formula lets people calculate roughly how huge that amount is.

Economics, in particular, is full of humanistic applications of differential equations. One of the most famous is the Black-Scholes equation, a differential equation that defines how much a stock option should be worth (and in particular, identifies the key aspects of such an option that control its price). Scholes won the Nobel Prize in 1997 for the development of this model, an indicator of its significance. But differential equations underly many other aspects of economics, including measures of risk, the search for equilibrium and stability, and changes in labor productivity.

6.6 Computational Methods of Solution

Much of the theory of differential equations centers around the the idea of finding a specific category of equation (like homogenous first-degree linear equations) and finding a specific method to solve them. Earlier in the chapter, we discussed trying to find "guidelines" for solving differential equations. It would be really nice if we could present a general framework where you simply turn the crank and the answer pops out, no matter how big or ugly the equations—an algorithmic method such as we use to solve long-division problems of any size.

Unfortunately, we can't do that. In fact, no one can. No one knows of such a framework, and it very probable that such a framework doesn't exist. Differential equations research is, for the reason, a rather *ad hoc* collection of different categories (such as first-order linear equations) and methods that work on specific categories for precisely this reason. But the categories that people look at tend to be the ones that arise in the real world. In other words, if you know enough categories and enough methods, you can solve most of the problems that come up, especially since the ones that come up often are the ones people have tried hardest to solve.

But there's an alternative approach—one that has become very

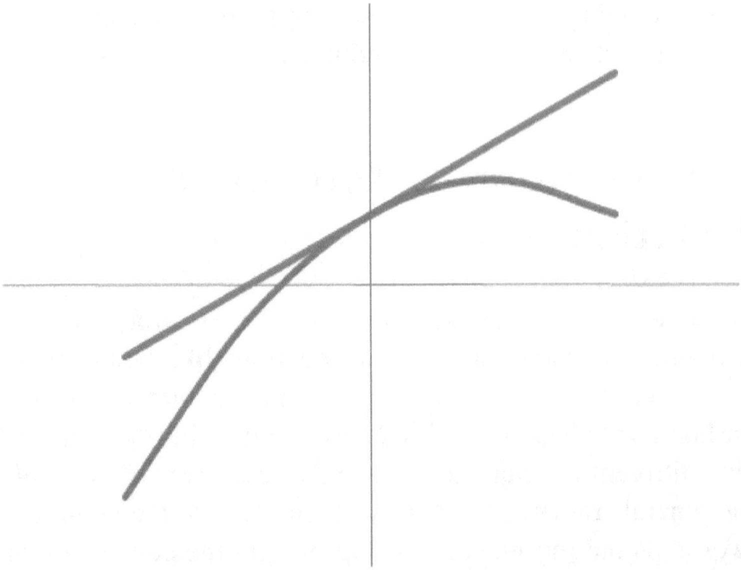

Figure 6.9: Using a first derivative to predict changes in a function

popular in the age of computing. Instead of trying to derive an exact formula for the family of solutions, why not use a computer to generate numeric solutions to differential equations instead? After all, the structure of a differential equation lends itself immediately to a physical model. You have a parameter t, which generates a function $y(t)$. At time 0, you know the value of the function $y(0)$, and if necessary the values $y'(0)$, $y''(0)$ and so forth. Using this information, you can compute the values of the highest derivatives, which tell you how each of the values you have will change.

For instance, if I know that $x(0) = 1$ and that $x'(0)$ also $= 1$, I know that at time 0-and-a-little bit, x will be a little bit more. We can go further, and use the actual value of the derivative to extrapolate a possible new value. For example, if the derivative is indeed 1, then a naive linear extrapolation would suggest that new values should be on the straight line (0,1), (1,2), (2,3), and so forth. (See figure 6.9.)

As the figure shows, this method is at best inexact. However, if the extrapolation is done over short enough distances, and the differential equation is well-enough behaved, we can get a good approximation of the behavior of the system over reasonably-sized intervals. Modern software systems (which are widely available) are often able to estimate the value of a differential equation to within one part in a billion or so, so unless you need truly phenomenal amounts of precision, computational methods will usually work well enough.

6.7 Chaos Theory

The parts of differential equations where you do need phenomenal precision is one of the areas of mathematics that has captured the imagination, if not the understanding, of the public at large. Chaos theory, in its essentials, is the study of the systems of differential equations where small changes in the problem specification produce huge differences in the solution.

6.7.1 Bifurcation

Normally—and what a wonderfully assumption-filled word that is—we do not expect small changes in a process to make big changes in the output. An extra splash of milk will not render the bread recipe inedible. If running 100m takes 15 seconds, running 101 meters should not take four hours. On the other hand, we've seen examples where a small difference in a differential equation can make a huge difference in the form of the output.

For example, equation 6.3, the Malthusian model of population growth (repeated here)

$$\frac{dP}{dt} = k \cdot P \tag{6.54}$$

has the solution $P = Ae^{kt}$. If $A > 0$ (and $k > 0$), then this solution is a steadily increasing function. If $A = 0$, then the function is constant 0 and never increases at all. If $A < 0$ (a situation that will probably not happen in actual population studies), then the population will be negative and will actually decrease over time. A similar argument can be made about k. Imagine our concern, then, when we perform an experiment to find A and get a result of 0! Any experiment will, of necessity, have a certain amount of potential error associated with it. Perhaps if we had gotten more data, we would have been able to measure the value of A at 0.0001, which means that the population will increase (slowly).

In real terms, we can think about this as a measure of the difference between "none" and "a few." One could, in a violent attempt to control the population of cane toads—an highly destructive invasive species—in Australia by killing all the cane toads. The population would then become (and remain) zero. But if I left even a single breeding pair of toads, that breeding pair could eventually repopulate the entire continent, and I'd have to start back from square one. In practical terms, though, how could I ever be sure that I had killed all the toads? How could I know that I hadn't missed some in a well-hidden swamp?

Other examples of bifurcation include the difference between overdamped and underdamped solutions to the oscillators generated by second-order equations; again, the behavior of the function will be qualitatively different with very small changes in the parameters. We've also seen how the annuity pricing model can give different behaviors; if the amount you take out is more the amount you earn, your balance will eventually go to zero. If the amount you take out is less than the amount you earn, you will get richer and richer while still having a steady income. (This is just a mathematical formalization of Wilkins Micawber's famous observation that "Annual income twenty pounds, annual expenditure nineteen nineteen six, result happiness. Annual income twenty pounds, annual expenditure twenty pounds ought and six, result misery.")

In a sense, we can characterize the "space" of possible Malthusian behavior by looking at all possible values of A and k, which would give us a two-dimensional space. We could also look the gross behavior of each particular solution, and we would find that regions of this space tend to display similar types of behavior; when A and k are both positive, the function grows without bound as an exponential, but when A is negative and k is positive.

In between is a thin line at $A = 0$ that acts like a continental divide

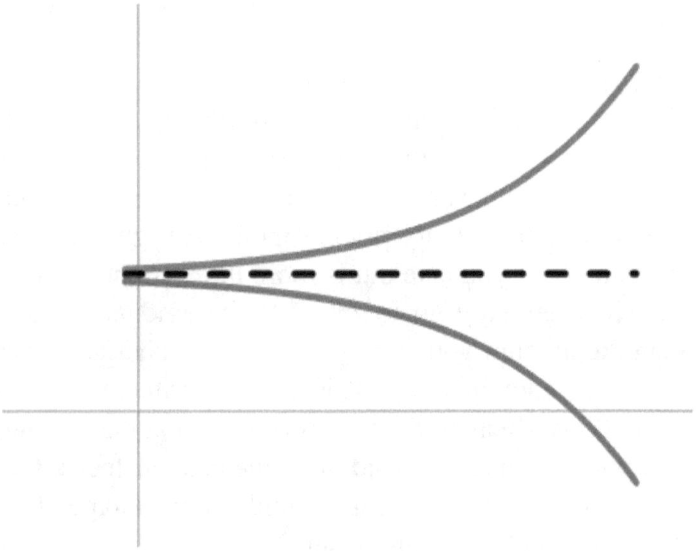

Figure 6.10: Two separate growth curves, depending on how much you withdraw compared to your income

to separate one region from another. This continental divide is a point where the function bifurcates (and where a single shilling can make the difference between happiness and misery).

6.7.2 Basins of Attraction and Strange Attractors

The usual term for these regions is "basins of attraction." Like a puddle forming at the bottom of a pit, the idea is that starting at any point in this basin will produce similar behavior. In the logistic model, the population capacity K is an example of a point attractor, the "lowest point" in such a basin. If the population is positive but less than K, it will tend to rise to K. If the population is larger than K, the population will fall until it reaches K. The number K thus "attracts" the population and will serve to stabilize any variations.

In addition to point attractors, it is possible to have stable but cyclic attractors. The earth, for example, is stable (more or less) in its orbit around the sun, but is not fixed to any one point. Cyclic, in this case, doesn't necessarily mean circular, just that the path through the space of numbers will eventually repeat itself. (Cometary orbits are far from circular, as is the boom-bust cycle of a typical predator-prey relationship, but they are still cyclic in this sense.)

The most unusual sort of attractor, and the one that has gotten the most attention, is the so-called **strange attractor**. A strange attractor is a path through space that is stable, but does not repeat itself. In practical terms, it is a basin of attraction with a very complicated boundary—one so complicated that we cannot apply a simple rule to see whether or not a given point is within its basin of attraction.

6.7.3 Chaos and the Lorenz Equations

This phenomenon was first discovered by the meteorologist Edward Lorenz (1917–1981) in 1961. As you might expect, there are a lot of differential equations involved in atmospheric science; the change in humidity is related to the temperature, but the temperature is related to the humidity, and so forth. Not only are there a lot of equations, but they all interlock, so it's difficult if not impossible to solve any of them by itself.

Lorenz was looking at the following system of equations:

$$\frac{dx}{dt} = -10x + 10y$$
$$\frac{dy}{dt} = 30x - y - xz$$
$$\frac{dz}{dt} = -3z + xy$$

These are not linear equations, since some of the variables are multiplied against each other (like xy). It is not practical to solve them directly, but they can be solved (more or less) using a computer. Lorenz found that this system of equations could be solved, but that it was incredibly sensitive to the initial conditions; changing a value from 0.506127 to 0.50600 (for convenience) would produce radically different results. They started out looking the same, but the tiny errors introduced by this minute change would rapidly multiply until the solution was not just inaccurate, but of a completely different form. This tiny change—in fact, almost any tiny change—seemed to cross into a different basin of attraction and follow a different attractor.

A similar problem had been identified earlier in the work of Poincaré on the three-body problem. The two-body problem (solved by Newton and his contemporaries) is a classic problem in mathematical astronomy; given two bodies, determine the paths that they will follow due to the mutual influence of gravity. However, if you add a third body into the mix, the math gets much more complicated; not only does the Sun have an effect on Jupiter and Jupiter on the Sun, but both planets also are affected by (and affect) Saturn. The effects of Saturn are relatively minor, but profound in the long-term. it turns out that the three-body problem also generates a system where small changes in the initial conditions can have a huge effect on the long-term results.

In theory, a well-behaved differential equation is completely deterministic. If you know the constants (exactly) and the initial conditions (exactly), you can determine (exactly) how the system will behave for as far into the future as you like to project. In practice, you can never know the constants exactly. You might ask, "What's the mass of Jupiter?" but can we know that to a millionth of a gram? A billionth of a gram? A trillionth? In such a system, any error or uncertainty in measurement will unavoidably be multiplied without limit in a long-term simulation of the system.

This, then, is why we say that the Earth's orbit is only more-or-less stable, because it is in a system involving other planets, and we cannot determine exactly what influences they will have. We can with relative confidence predict that the Earth's orbit will not change much in the next hundred or thousand years, but we can say little about the next ten million or ten billion. We simply don't know Jupiter's mass accurately enough—and will never know it accurately enough. And so eventually Earth's orbit is likely to change radically under the influence of the other planets, but we can't say why, how, or when exactly.

This helps to explain why long-term weather prediction is so difficult and the predictions so inaccurate. The temperature affects the pressure affects the wind affect the rainfall. However, a person lighting a cigarette might make a tiny change (that we didn't know about) in the local temperature, and that tiny change will be multiplied in the weather system, and a month from now, the weather could be almost anything. No wonder such systems were dubbed "chaotic."

This is also the original source and the meaning of the term **butterfly effect**. A butterfly flaps its wings somewhere, and changes the wind patterns from what our computers say they are. Our computers can't track the wind accurately enough, but this slight change in the initial conditions is enough, eventually, to have a profound change on the overall weather.

Of course, the idea that small things can have huge effects is not new; as a popular proverb has it, "great oaks from small acorns grow." We like this famous poem about Richard III even better:

For want of a nail the shoe was lost.
For want of a shoe the horse was lost.
For want of a horse the rider was lost.

For want of a rider the battle was lost.
For want of a battle the kingdom was lost.
And all for the want of a horseshoe nail.

Chaos theory has joined Gödel's ideas about "incompleteness" and Heisenberg's ideas about "uncertainty" as one of those abstruse mathematical ideas that has managed to capture the popular imagination. As with earlier instances, the emphasis is often placed on what mathematics (or physics, or computers) cannot do. But in all of these cases—and especially in the case of chaos theory—the insights gleaned from advanced study of these phenomena have tended to be highly productive. Understanding where chaos appears has led to enourmous insights into the nature of highly complex, dynamic systems, ranging from ecologies to economies, and it has opened doors to interdisciplinary collaborations across widely divergent disciplines. Whenever we contemplate such matters as influence, genre, memes, and cultural change, we wonder if humanists might add further insights to this important area.

6.8 Further Reading

The book *Differential Equations*, by Polking, Boggess, and Arnold[7] is excellent; indeed, we have taken many examples and much of the discussion framing the examples from this book.

---

[7]Prentice-Hall, 2001

Some Final Thoughts

> Assaye in myn absence
> This disciplyne and this
> crafty science.
>
> Geoffrey Chaucer
> *The Canterbury Tales*

Well, there you have it.

In the introduction, we claimed that our purpose in writing this book was eminently practical. We intended to write a primer on advanced mathematics for humanists, and further maintained that having a background in the conceptual foundations of mathematical reasoning and its sometimes hieratic notation was the key to facilitating further research on data mining, geo-spatial analysis, computational linguistics, programming, image analysis, and the various other technical endeavors that make up what is now called the Digital Humanities.

We hope we've achieved that goal, but perhaps it's time to come clean and put forth some of our loftier motives.

University administrators are fond of the term "interdisciplinarity," and so are we. After all, we've both built scholarly careers that are fundamentally based on a desire to look over the fence, and we're hardly alone. For all that, we are aware that transcending one's own disciplinary training is, at some level, impossible. If you are (like one of us) an English professor, you have inherited not simply a list of books read and classes taken, but the tacit assumptions and values of an entire community. If you are (like the other of us) a professor of Mathematics and Computer Science, you have likewise inherited an entire worldview about what kinds of problems there are and what kinds of problems are worth investigating. Exchanging one kind of heritage for another is no easy a matter, and in the end, time is perhaps the least significant barrier. The living intellectual communities in which a discipline exists are surrounded by vast bureaucracies, rules, publishing regimes, tenure requirements, and peer-review systems that very often exhort one to get off their land (or stay on it).

We are aware, of course, of many successful attempts to venture forth. The biologist who gains facility with computation becomes a "computational biologist;" the historian who undertakes deep study of scientific matters becomes an "historian of science;" the linguist who ventures into biology becomes an "evolutionary linguist." But no sooner have these "interdisciplines" been born, than the cycle begins again. Departments and programs are created, students are taught, journals are established, and before long these once anxious pairings become just another walled garden. When that happens, those who pioneered the venture feel a sense of well-earned pride, but may also feel as if the giddy days of first contact have subsided—and with them, a sense of intellectual excitement and vitality.

One of the most difficult matters we've encountered in writing this book is getting the "tone" right. It is one thing for a curious intellectual to venture tentatively into another discipline; it is another thing entirely to suggest—as we implicitly do—that what one discipline knows, another should. In the end, we've spoken the only way we know how, with what we hope is the right mixture of humor and seriousness, boldness and humility. But our real goal in doing this was perhaps to evoke the ancient meaning of the word "discipline," which has its roots in a distinction between the property of the scholar or "doctor" *(doctrina)* and his or her practice *(disciplina)*. Our intention was not so much to create new kinds of scholars and scholarly categories, but to suggest new ways of conducting one's intellectual life.

It would have been possible to write a book in which we move through each mathematical subject and apply it to some humanities problem: topology for the GIS practitioner, predicate logic for the student of language, graph theory for the social network analyst. By now, it should be clear that we did not write that book—not because such a book couldn't be written, but because we felt that to do so would be to break faith with what we regard as the truly momentous *disciplina* of the "interdisciple." The professor of French literature who ventures, against all outside pressure, into differential equations is taking a bold, uncertain leap. We have no idea what will come of that. She has no idea what will come of that. We are certain that something truly great will come of that.

And what one fool can do, another can.

A Quick Review of Secondary-School Algebra

In several places in this book we invoke some elementary rule of mathematical manipulation without much comment—often by saying that "high school algebra shows" that something or other is true. We're referring, of course, to something like the basic "grammar" of mathematics, which most of us acquire prior to entering college.

We are aware, of course, that that moment of preparation may be a quite distant memory for many of our readers, and so we offer this appendix as a brief and brisk re-introduction to that basic set of mathematical rules and techniques. To be very explicit: this summary covers math below university level (what education professionals generally call "secondary school"). In the USA or Australia, this would be high school, while in England this would be GCSE-level. Since, in many cases, this could be material that you were first exposed to at the tender age of twelve or so, you've likely had lots and lots of time to forget some of the details. Have no fear; it will all come back to you in short order.

Algebra, in the context of higher mathematics, is a much broader and grander subject, which we discuss in Chapter 2. Nearly every chapter in this book, however, assumes familiarity of what we set forth here.

1 Arithmetic Expressions

An **expression** is any string of mathematical symbols that has a value. 7 is an expression (a very simple one), as is $3+4$, as is $(\sqrt{100}+4)/2$ (though, as it happens, they all have the same value: 7). In most circumstances, we would rather work with easy expressions like 7, so we spend a lot of time simplifying expressions. The idea is that we can replace any (complicated) expression by a simpler expression that has the same value, and there are a number of rules that have to be thoroughly absorbed in order to undertake the usually quite necessary process of making things simpler.

- $a+b=b+a$
- $ab=ba$

  There are a lot of ways to represent the idea of multiplying a by b; sometimes it's written $a \times b$, sometimes $a \cdot b$, sometimes just ab. They all have precisely the same meaning.

  It doesn't matter in what order we undertake addition and multiplication; it does matter for subtraction and division, since $3-2$ is obviously different from $2-3$.

- $(a+b)+c=a+(b+c)$
- $(ab)c)=a(bc)$

Similarly, it doesn't matter which two numbers we add/multiply first when we have a bunch of them to work with.

- $a(b+c) = ab + ac$

  Multiplying a sum is the same as adding products. This implies an **order of operations**; e.g., multiplication happens before addition unless you use parentheses to override that rule. If you dig deep, you may recall the most famous mnemonic in all of mathematics: "Please Excuse My Dear Aunt Sally" (e.g. Parenthesis, Exponents, Multiplication, Division, Addition, Subtraction).

- $-1(a) = -a$

- $-(-a) = a$

- $-(a-b) = (b-a) = -a+b$

  Subtraction is really just adding numbers "times negative one" so the stuff above also works.

- $a \div b = a/b = \frac{a}{b} = a(\frac{1}{b})$

- $a = \frac{a}{1}$

- $(\frac{a}{b})(\frac{c}{d}) = \frac{ac}{bd}$

- $(\frac{a}{b}) \div (\frac{c}{d}) = (\frac{a}{b})(\frac{d}{c}) = \frac{ad}{bc}$

- $(\frac{a}{b}) + (\frac{c}{b}) = \frac{a+c}{b}$

- $(\frac{a}{b}) + (\frac{c}{d}) = \frac{ad+bc}{bd}$

  Fractions and division behave exactly as they did when we first suffered through them, including that business about "common denominators." You'll also recall that division by zero is illegal (or, to speak more properly, "undefined").

- $(\frac{ca}{cb}) = (\frac{a}{b})$

  You can reduce fractions by dividing out common factors (as long as c isn't zero).

- $(a^b)(a^c) = a^{b+c}$

- $(a^b)^c = a^{bc}$

- $(ab)^c = a^c b^c$

- $a^{-b} = \frac{1}{a^b}$

- $a^0 = 1$ as long as $a \neq 0$

  Exponents are used to express repeated multiplication ($a^2 = a \cdot a$ and so forth).

- $\sqrt{a} = b$ if and only if $b^2 = a$

- $\sqrt[n]{a} = b$ if and only if $b^n = a$

  Radicals are used to signify the reverse of exponentation; taking the square root (or any even root, such as a fourth root or a sixth root) of a negative number is not generally recommended.

We can assure you that those rules were every bit as boring to type as they were to read, but they consititute the irrefragible rules of arithmetic. Without complete mastery of them, it becomes very difficult to solve most problems in mathematics. Once fully internalized, though, solving equations becomes a mostly mechanical matter, as when one needs to convert 40 degrees Fahrenheit into degrees Celsius:

$$\frac{9}{5}(40)+32$$
$$\frac{9}{5}\frac{40}{1}+32$$
$$\frac{360}{5}+32$$
$$72+32$$
$$104$$

What's more, these basic rules contain some deep intellectual treasures, which, as we noted, are treated in Chapter 2.

2 Algebraic expressions

Things get a little bit more interesting when we introduce truly algebraic expressions—expressions involving variables like x and y. The basic rule is that we can only add and subtract "like" terms, which is to say, terms with the same variables. So using the distributive law above ($a(b+c)=ab+ac$) we know that $2x^2+5x^2=(2+5)x^2=7x^2$. Similarly, $(x^3-2x^2+5x-2)+(x^2-2x-1)=(1+0)x^3+(-2+1)x^2+(5-2)x+(-2-1)=x^3-x^2-3x-3$.

To multiply algebraic expressions, we apply the distributive law several times:

$$(a+b)(c+d)$$
$$a(c+d)+b(c+d)$$
$$ac+ad+bc+bd$$

There are two basic forms that are easy to work with. One is the simple polynomial, where the expression is written as a sum of decreasing powers of a single variable, like $3x^3+2x^2+7x+1$. The second is as the product of **linear** terms (each term having only a single variable raised to the power 1), like $(3x-2)(x+7)(4x+1)$. The process of getting from the first basic form to the second is called **factoring**, and it is an important, if tedious operation (which we won't cover here). Similarly, dividing one algebraic expression by another can be very difficult; we won't worry about it for now.

3 Equations

An **equation** is what you get when you assert that two expressions are equal. For example, if we say $3x+5=14$, we are saying that the value of 14 is the same as the value of $3x+5$. Of course, this isn't always true; if the value of x were 0, then $3x+5$ would be 5, not 14. But by solving the equation, we can figure out what the value of x must be, *if* the expression $3x+5$ were actually equal to 14.

To do this, we make any changes we feel like making, so long as we make the same changes on both sides of the equation (while carefully avoiding illegal operations like division by zero). Usually, one tries to group all of the terms containing x on one side, everything else on the other, and then divide. For example:

$$\begin{aligned} 3x+5 &= 14 \\ 3x+5-5 &= 14-5 \\ 3x &= 9 \\ \frac{3x}{3} &= \frac{9}{3} \\ x &= 3 \end{aligned}$$

so if $3x+5=15$ it must be the case that $x=3$.

Things get a little more tricky if you have higher powers of x involved. An equation like $x^2-4=5$ isn't too bad until you have to take the square root:

$$\begin{aligned} x^2-4 &= 5 \\ x^2-4+4 &= 5+4 \\ x^2 &= 9 \\ \sqrt{x^2} &= \sqrt{9} \\ x &= \sqrt{9} \end{aligned}$$

but remember that $\sqrt{9}$ actually has two different values, $+3$ and -3. So x can be (must be) either of those values.

Things get considerably more difficult if you have many different terms involving powers of x. This is where factoring comes in. Consider the equation $4x^2 - 3x + 5 = 6$. It is fairly simple matter to isolate the terms involving x on one side, but it's hard to solve directly. Product-of-terms notation can help a lot. Let's do some of the preliminary work:

$$\begin{aligned} 4x^2 - 3x + 5 &= 6 \\ 4x^2 - 3x + 5 - 6 &= 6 - 6 \\ 4x^2 - 3x - 1 &= 0 \end{aligned}$$

At this point, factoring the polynomial into product-of-terms form yields

$$(4x+1)(x-1) = 0 \qquad (1)$$

If you multiply two nonzero numbers together, you get a nonzero number. Another way of saying the same thing is that if one number times another equals zero (as in equation 1), at least one of the numbers must itself be zero. So if $(4x+1)(x-1) = 0$, then *either* $4x+1 = 0$ *or* $x-1 = 0$.

This gives us two equations to solve separately, and which will render two different values for x:

$$\begin{aligned}
4x+1 &= 0 \\
4x+1-1 &= 0-1 \\
4x &= -1 \\
\frac{4x}{4} &= \frac{-1}{4} \\
x &= -\frac{1}{4}
\end{aligned}$$

$$\begin{aligned}
x-1 &= 0 \\
x-1+1 &= 0+1 \\
x &= 1
\end{aligned}$$

so x is either $-\frac{1}{4}$ or 1. Harder equations involving higher powers of x, or involving more variables than just one, can be solved similarly.

4 Inequalities

An **inequality** is like an equation, but it's what you get when you assert that one thing is more or less than another. For example, if you need to buy a dozen eggs but are unwilling to pay more than \$3.00 in total, that doesn't mean you wouldn't take them if they were given away for free. What's the *maximum* price per egg you would be willing to pay?

Well, if x is the price of a single egg, the price of a dozen is $12x$. Since that's the maximum price, the price you pay for twelve needs to be less than or equal to ($\leq$) \$3.00.

$$\begin{aligned} 12x &\leq 3.00 \\ \frac{12x}{12} &\leq \frac{3.00}{12} \\ x &\leq \frac{1}{4} \end{aligned}$$

So you shouldn't pay more than $\$\frac{1}{4}$ or 25 cents per egg. Solving inequalities is just like solving equations, with one added trick. Again, you can do anything you like as long as you do the same thing on both sides—*but,* you have to switch the inequality (from $<$ to $>$ or vice versa) if you multiply by a negative number.

Which makes sense. If $x > 5$ (maybe x is 10), then $-x < -5$ (because $-x$ would be -10).

5 Story Problems and Mathematical Models

The other important part about high school algebra is learning how to develop mathematical models from "story problems." The previous section had a simple example involving eggs and the difference between price per dozen and price per single egg. But these types of problems arise in many (quite abstruse and difficult) contexts.

Solving story problems is a skill, and one that takes practice to master. But the steps involved in solving them are easy to express in general terms:

1. Identify what the problem is asking you to solve, and use a simple variable like x to represent it. Usually you'll know what the problem is asking for by looking at the question posed at the end of the problem (e.g. *"What's the maximum price per egg you would be willing to pay?"* We called that x.)

2. Figure out what information you are given (in this case, the maximum price per dozen eggs and, implicitly, the number of eggs in a dozen). Therefore we must...

3. ...find a way to represent the relationship between the information you have and the answer you want as an equation or an inequality. If one egg costs x, a dozen eggs cost $12x$.

4. Plug the values you have into the equation or inequality, and then solve it. This will give you the value for x (your answer).

5. Check your work, because everyone makes mistakes in math sometimes. We've been doing this stuff professionally for decades, and are still fully capable of messing up an apparently straightforward problem involving eggs.

A couple of examples will illustrate how this is done. Let's suppose that it takes twelve hours to dig a hole two meters long, three meters wide, and a meter deep. I need a hole dug that is four meters long, two meters wide, and two meters deep. How long will it take to dig the (larger) hole?

1. The problem is asking how long it will take to dig a hole of a specified size. As a working assumption, the time it takes to

dig a hole depends on how big the hole is—a hole twice as big takes twice as long.

2. You are given the time it takes to dig a hole of a (different) size.

3. Using this information, you should be able to figure out the rate at which holes are dug (the number of cubic meters dug per hour); it's just the volume of the smaller hole divided by the time it takes to dig that hole. If holes are dug at the same rate, the volume of the second hole divided by the time of the second hole will be the same value.

   The formula for volume is just length times width times height. So the smaller hole has a volume of $(2)(3)(1)$ and the larger hole has a volume of $(4)(2)(2)$.

4. Now we have our equation and can plug in our values:

$$
\begin{aligned}
\frac{(2)(3)(1)}{12} &= \frac{(4)(2)(2)}{x} \\
\frac{6}{12} &= \frac{16}{x} \\
\frac{(12x)(6)}{12} &= \frac{(12x)(16)}{x} \\
6x &= (12)(16) = 192 \\
\frac{6x}{6} &= \frac{192}{6} \\
x &= 32
\end{aligned}
$$

so it should take 32 hours to dig my hole.

5. Does this answer make sense, or did we embarrass ourselves? Well, the volume of the first hole is 6 cubic meters, and the volume of the second is 16 cubic meters. It takes twelve hours to dig the first hole, or two hours per cubic meter. It takes thirty-two hours to dig sixteen cubic meters, or (again) two hours per cubic meters. So the answer checks out.

Another example comes to us from classical Greece; the mathematician Diophantus of Alexandria (3rd century BCE) has given his name to a classic story problem. According to an anthology of number games, the following served as Diophantus' epitaph:

> Here lies Diophantus," the wonder behold.
> Through art algebraic, the stone tells how old:
> "God gave him his boyhood one-sixth of his life,
> One twelfth more as youth while whiskers grew rife;
> And then yet one-seventh ere marriage begun;
> In five years there came a bouncing new son.
> Alas, the dear child of master and sage
> After attaining half the measure of his father's life chill fate took him. After consoling his fate by the science of numbers for four years, he ended his life.

. . . and judging from that last line, it would appear that his interest in poetic meter was likewise cut short. But the question is this: How long did Diophantus live?

1. We want to know how old Diophantus was when he died (or alternatively, how long he lived). We will call that x.

2. We know that Diophantus' youth was $\frac{1}{6}$ of his life ($\frac{1}{6}x$). After another $\frac{1}{12}x$ he grew a beard. After another $\frac{1}{7}x$ he married. Five years later, he had a son, who lived $\frac{1}{2}x$ years. After his son died, Diophantus died 4 years later.

3. So Diophantus lived
$$\frac{1}{6}x+\frac{1}{12}x+\frac{1}{7}x+5+\frac{1}{2}x+4$$
years in total.

4. We thus have two equivalent expressions for how long Diophantus lived. He lived x years, and he also lived $\frac{x}{6}+\frac{x}{12}+\frac{x}{7}+5+\frac{x}{2}+4$ years. Setting these equal to each other (since they're both the same age) gives us the equation:

$$\frac{x}{6}+\frac{x}{12}+\frac{x}{7}+5+\frac{x}{2}+4 = x \qquad (2)$$
$$\frac{14x}{84}+\frac{7x}{84}+\frac{12x}{84}+\frac{42x}{84}-\frac{84x}{84} = -9 \qquad (3)$$
$$\frac{-9x}{84} = -9 \qquad (4)$$
$$(-9)x = (-9)(84) \qquad (5)$$
$$x=84 \qquad (6)$$

so Diophantus lived to be 84.

5. And, checking the math, it hangs together. He was a youth until age 14, grew a beard at age 21, and married at 33. His son was born when he was 38, lived until he was 80, and then Diophantus died four years later at age 84.

As a final example, consider a poor conference organizer who has to get 330 guests from the conference hall to the banquet. He can rent buses that hold 50 people each, but needs to figure out how many buses will be needed.

1. He needs to know how many buses he needs to rent. Call this x.

2. Each bus holds 50 people, and he has 330 people to move.

3. ... so x buses will hold $50x$ people. We can thus set up a simple equation $50x = 330$.

4. Solving for x, we get $x = \frac{330}{5}$ or $x = 6.6$, so we need 6.6 buses.

5. But does this make sense? Not really. Where the devil do you go to rent six-tenths of a bus? Checking our work: Since buses need to be rented as whole buses, he needs to rent *seven* buses, which have a total capacity of 350 people. Renting only 6 buses would only allow 300 people to get there, which isn't enough.

Alternatively, we could have set this up as an inequality (since he needs to transport *at least* 330 people. This would have given us the inequality $50x > 330$ or $x > 6.6$, which again suggests a need to rent seven buses.

We'll admit that this is a matter of common sense, but it's an object lesson that becomes remarkably important when dealing with more complicated matters (including such matters as physics and economics). If the equations tell you that the Second Law of Thermodynamics is invalid or that an investment will yield 100,000 times its value, it's likely that the work is incorrect. It may also, however, contain an underlying assumption that doesn't "make sense" to begin with. Just because you can set up and solve an equation doesn't mean that you have to accept whatever gibberish comes out of the math.

6 Further Reading

The home-schooling industry, the university-preparation industry, the get-a-GED industry, and the general publishing industry's fondness for dummies, idiots, and other humiliated populations, has yielded a number of positively superb books on algebra and arithmetic for those who need more tutorial explanations, a more detailed review, or simply more examples. Needless to say, most of the books with the word "algebra" in the title (or table of contents) are about high school algebra, and not the more abstract matters discussed in Chapter 2. Similarly, anything with the world "precalculus" in the title will cover the upper half of high school algebra as well as more advanced material like exponential, trigonometric, or logarithmic equations (equations like $10^x = 2x$). Even the most confident among us will be well served by having a short, basic reference on these matters—right next to this one.

Bibliography

[1] Aristotle. *The Complete Works of Aristotle*, chapter Physics. Princeton UP, Princeton, 1984.

[2] E. T Bell. *Men of Mathematics: The Lives and Achievements of the Great Mathematicians from Zeno to Poincaré*. Simon-Touchstone, New York, 1937.

[3] Mohammed Ben Musa (Muḥammad ibn Mūsā al Khwārizmī). *The Algebra of Mohammed ben Musa*. London, 1831.

[4] Euclid. *The Thirteen Books of the Elements*. Dover, New York, 1956.

[5] G. H. Hardy. *A Mathematician's Apology*. Cambridge UP, Cambridge, 1940 edition, 1967.

[6] David Hilbert. *From Kant to Hilbert: A Source Book in the Foundations of Mathematics*, volume 2, chapter Mathematical Problems, pages 1096–1104. Oxford UP, New York, 2007.

[7] Alfred Jarry. *Selected Works of Alfred Jarry*. Grove, New York, 1965.

[8] Robert Kanigel. *The Man Who Knew Infinity: A Life of the Genius Ramanujan*. Washington Square Press, New York, 1991.

[9] Immanuel Kant. *Critique of Pure Reason*. Cambridge UP, Cambridge, 1998.

[10] Bertrand Russell. *Autobiography*. Routledge, London, 1967.

[11] Richard J. Udry. The effect of the great blackout of 1865 on births in new york city. *Demography*, 7.3, 1970.

www.ingramcontent.com/pod-product-compliance
Lightning Source LLC
Chambersburg PA
CBHW021827220426
43663CB00005B/151
9781609621117